Jacob Davis Babcock Stillman, Eadweard Muybridge, Leland Stanford

The Horse in Motion

Jacob Davis Babcock Stillman, Eadweard Muybridge, Leland Stanford

The Horse in Motion

ISBN/EAN: 9783337236236

Printed in Europe, USA, Canada, Australia, Japan

Cover: Foto ©berggeist007 / pixelio.de

More available books at **www.hansebooks.com**

THE

HORSE IN MOTION

AS SHOWN BY INSTANTANEOUS PHOTOGRAPHY

WITH A STUDY ON ANIMAL MECHANICS

FOUNDED ON ANATOMY AND THE REVELATIONS
OF THE CAMERA

IN WHICH IS DEMONSTRATED THE THEORY OF QUADRUPEDAL LOCOMOTION

BY J. D. B. STILLMAN, A.M., M.D.

———————— .

EXECUTED AND PUBLISHED UNDER THE AUSPICES OF

LELAND STANFORD

BOSTON
JAMES R. OSGOOD AND COMPANY
1882

UNIVERSITY PRESS:
JOHN WILSON AND SON, CAMBRIDGE.

PREFACE.

I HAVE for a long time entertained the opinion that the accepted theory of the relative positions of the feet of horses in rapid motion was erroneous. I also believed that the camera could be utilized to demonstrate that fact, and by instantaneous pictures show the actual position of the limbs at each instant of the stride. Under this conviction I employed Mr. MUYBRIDGE, a very skilful photographer, to institute a series of experiments to that end. Beginning with one, the number of cameras was afterwards increased to twenty-four, by which means as many views were taken of the progressive movements of the horse. The time occupied in taking each of these views is calculated to be not more than the five-thousandth part of a second. The method adopted is described in the Appendix to this volume.

When these experiments were made it was not contemplated to publish the results; but the facts revealed seemed so important that I determined to have a careful analysis made of them. For this purpose it was necessary to review the whole subject of the locomotive machinery of the horse. I employed Dr. J. D. B. STILLMAN, whom I believed to be capable of the undertaking. The result has been that much instructive information on the mechanism of the horse has been revealed, which is believed to be new and of sufficient importance to be preserved and published.

The HORSE IN MOTION is the title chosen for the book; for the reason that it was the interest felt in the action of that animal that led to the experiments, the results of which are here published, though the interest wakened led to similar investigations on the paces and movements of other animals. It will be seen that the same law governs the movements of most other quadrupeds, and it must be determined by their anatomical structure.

The facts demonstrated cannot fail, it would seem, to modify the opinions generally entertained by many, and, as they become more generally known, to have their influence on art.

LELAND STANFORD.

PALO ALTO FARM, CALIFORNIA,
1881.

CONTENTS.

CHAPTER I.

CHAPTER II.

CHAPTER III.

CHAPTER IV.

LIST OF PLATES.

———•———

THE HORSE IN MOTION.

CHAPTER I.

THE Horse, of all animals, holds the most important relations to the human family. Though the earliest traces of his existence on the globe are found as fossils in North America, as an historical character he is traced to Central Asia with the Caucasian race. There was no representative of the race living in America at the time of the discovery of the New World, but it was introduced by Columbus and his followers, and its descendants became feral on the Prairies of North and the Pampas of South America. They were undoubtedly of Arabian stock, through the Moors; small, active, and hardy. Their descendants were very numerous in what were the northern provinces of Mexico, previous to the invasion of Texas.

The genera were well represented in Africa and the deserts of Arabia, but we have no evidence that the historic horse was known in Africa before the time of Rameses the Great. in the Eighteenth Dynasty, after the wars with the Persians. Nowhere in all the temples and tombs of Memphis, Sais, Abydos, of the First Empire, is there a sculpture that could lead us to infer that the horse was known to the Egyptians of that early age. There are no sculptures in India older than the dawn of Buddh, or about five centuries before our era. The oldest written account of the horse is found in the book of Job, and that is a very spirited description of a war-horse; and it is probable that that is the oldest of the sacred writings of the Hebrews, though there is no clew to the date or origin of that curious production.

Though the relative importance of the horse as a factor in the progress of civilization has been reduced by the introduction of steam in our century, it cannot be forgotten that he has been the constant companion of the Caucasian race in all its migrations, an indispensable ally in all its conquests, and the most efficient agent of its civilization. We have no history that is not interwoven with his; and if by some sudden cataclysm he should be eliminated, we should then be made to realize how indispensable he still is to our business and pleasure. Whatever concerns him will never cease to interest mankind.

The interest in the paces of the horse is not new: it had engaged the attention of philosophers from ancient times.   Aristotle, the father of Philosophy, thought it not unworthy his investigation; but with all other rational questions, it was lost to human thought during the long reign of religious bigotry. When the intellects of men were again set free, and Science woke from her slumber, Anatomy was studied and taught in the schools, and attention became directed to that of our subject; but even Borelli, who wrote about two hundred years ago, and published the work on Animal Mechanics that most later writers have drawn upon, thought it necessary that he should not confound flesh and muscle. Vital force was as yet unknown, and all treated the subject as a physical science, and deduced its laws from the motions of the pendulum, and mathematically formulated them.

Two brothers, named Weber, who are quoted much by the author of "Animal Motion," in the "Encyclopædia of Anatomy and Physiology," followed Borelli on the purely physical theory of Animal Motion.

Professor Marcy has contributed the result of many laborious and painstaking experiments on the slow paces, by means of apparatus attached to the feet, and connected by elastic tubes with registers in the hands of the rider. This apparatus would determine the force of the footfalls and time of pressure, and by the system of notation a chart could be made of the paces. But it failed to interpret the paces correctly, or furnish the basis of a theory of quadrupedal locomotion. The importance of the subject had been fully appreciated by him, as appears in the following quotation from his work on Animal Mechanics: "There is scarcely any branch of animal mechanics which has given rise to more labor and greater controversy than the question of the paces of the horse. The subject is of great importance to a large number of persons engaged in special pursuits, but its extreme complexity has caused interminable discussion. Any one who proposed at the present time to write a treatise on the paces of the horse would have to discuss many different opinions put forward by a great number of authors."

Bishop, the author of the article on "Animal Motion" in the "Encyclopædia of Anatomy and Physiology," says: "The study of the mechanism of which the locomotive organs of animals is composed, of the laws by which their progression is accomplished, and of the vital force which they expend in propelling the body from one point in space to another with different velocities, serves to interest alike the anatomist and the physiologist, the artist and the mechanician. Ignorance of these laws has been productive of grotesque delineations of the human figure as well as of the lower animals, when represented in motion. We have abundant evidence of this in the productions of painters and sculptors, both of the ancient and modern world."

The difficulty in this, as in many controverted questions, is to determine the facts; and the facts have been most difficult to obtain. It seems to many unaccountable, that the horse, whose movements are so open, should play such a *léger-de-pied* as to deceive all eyes, and give rise to controversies as earnest as did the colors of the chameleon in the fable. All attempts hitherto made to analyze these movements have failed, for it is not possible for the eye to distinguish them; or rather, to state the case more accurately, the mind is unable to distinguish the impressions conveyed to it through the eye.

Controversies were going on to the last as to which foot was advanced first in the trot; whether the toe or heel first touched the ground; whether in a gallop the legs were stretched out fore and aft, or the knees were flexed. All were dabbling in the shallow waters of a sea whose depths there was no known method of exploring, and

artists of all degrees fell into the false and conventional manner of
representing animals in rapid motion, as untrue as were the Greek
conceptions on the subject thirty centuries ago. To understand how
little progress has been made in modern times, it is only necessary to
look at the productions of the best animal painters of our day.

Why is it that there have been such widely different interpre-
tations of these movements from the time of Aristotle down to the
present? These positions, as well as all others that have been represented, are proved by the unerring finger of light to be incorrect;
as mechanical anatomy, had it been properly consulted, would have
demonstrated to be impossible.

It is difficult at a glance to conceive how the eye could be so
deceived; but a little consideration of the physiology of that organ
will teach us that no dependence can be placed on it to interpret the
motion of an object moving irregularly, even at a comparatively slow
rate of speed.

It has been shown that the retina of the eye is capable of receiving a distinct image of an object in an almost inconceivably short
space of time, as that of the flash of an electric spark, or a millionth
part of a second, and that the impression remains for the space of a
third to a seventh of a second, according to the experiments of D'Arcy
and Plateau; and the mind is incapable of distinguishing between
the first impression and the last made during that space of time, and
the images run together and are confused. A familiar illustration of
this phenomenon is furnished by the spokes of a wheel in motion; yet
these spokes will appear stationary, if, revolving in the dark, they are
suddenly illuminated by an electric flash; or if the end of a stick be
ignited, and moved rapidly, a continuous line of fire will appear. Here
there is a continuous line of impressions made upon the retina, and so
conveyed to the mind. The same is true of the auditory nerve; when
vibrations of air are too rapid, they are heard, but not distinguished.
The reader may ask why it is that the artists of all time, with the full
accord of all men, — and our own eyes confirm the tradition, — represent
the horse in galloping as extending his feet to the utmost, as seen in
all the pictures of horses racing. My answer is this: We now know

that it is not true that a horse ever did put himself in the position portrayed by the best artists; and the explanation that I have to offer is, that in the gallop the horse always moves his feet alternately, and to the same extent; at the limit of extension there is a change of direction given to them, and their image dwells longer upon the retina, and the impressions are more lasting than of the intermediate and more rapid movements which the mind is unable to distinguish any more than the order in which they are made.

The ear has been relied upon to determine by the rhythm of the footfalls the order in which the feet strike the ground; and bells have been attached to the feet, each giving a different sound. Others have studied the footprints, and the feet have been differently shod to distinguish the impression made by each foot upon the ground.

The study of the mechanical anatomy of the horse is a necessity in order to a proper understanding of the forces employed and their combined action. This necessity has now become more imperative, as the action is better understood from the revelations of the camera.

All the systematic works on the anatomy of the horse have followed the plan of those on human anatomy, and apparently for the same purpose, namely, the intelligent treatment of the diseases and accidents to which horses, as well as men, are liable, while the action and relation of the machine, as such, have been treated as of secondary importance or altogether neglected. It has not been possible to study the action of the muscles singly without falling into errors; the correlation of all of them is necessary to the understanding of any one. It is to this cause that so many errors and contradictions found in all authorities that have been consulted are to be ascribed. Indeed, how was it possible that it should have been otherwise, so long as it was not known what those actions were?

The progressive motions of a quadruped, which must be considered as a unit, are very complex; when so studied it will be found that all the parts are mutually dependent, that the forces employed are compound and often indirect, and that the compensation of one indirect action may be found quite remote. When thus considered it will be found that the horse in motion is as perfectly harmonious in the dis-

play of his forces and their balance as a steam hammer, which may be
adjusted to a force sufficient to forge a shaft for an ocean steamer or
to crack a nut.

It cannot be expected that many of those persons who are interested in the movements of the horse will be familiar with the anatomical terms necessary to be used in the description of the simplest
motion, and it cannot be made intelligible without them; much less
can it be expected that one will be able to comprehend a full stride
from any analysis that can be given without such knowledge.

The writer thinks himself warranted in the assertion that the correct
interpretation of the mechanical action of the horse cannot be obtained
from any existing work. It is very desirable that it should be understood by every one who is interested in his achievements, and by artists
as well. To facilitate this study, technical terms will be omitted as far
as possible, and where they are employed they will be accompanied by
popular ones as far as they are known.   .

One of the sources of difficulty to the non-professional student is
the distinctive names given to different tissues whose mechanical
function is the same. Whether a muscle has its termination in facia
aponeurosis or at the bone on which it acts, either directly or indirectly, may be important to the anatomist or surgeon; but to those
who desire to understand the mechanical action it is a matter of indif-
ference, very perplexing, and a fatal bar to the comprehension of the
subject; to such it is of little consequence whether the action is direct
by muscular attachments to bones, or indirect through facia or other
fibrous tissue. In all cases I shall use such terms as will most correctly give my meaning in the interpretation of their action.

Another source of confusion in the study of the muscles of motion in quadrupeds is the conflicting names given to them. When,
on the restoration of the cultivation of science, comparative anatomy
began to attract the attention of naturalists, human anatomy had
already received much attention, and names had been bestowed upon
all the principal organs. Some of them were purely fanciful; others
were based on their resemblance to other objects. The muscles were
often named from their supposed function, or their correspondence to

muscles found in the human body. This last has been the most fruit-
ful source of confusion, and the mind of the student is constantly
biassed by this correspondence of names to muscles that do not have
corresponding functions. It may be taken for granted that organs
have the same diversity of form in man and animals as there is
diversity of function, and in each the organisms are just such as best
serve the offices which they were designed to perform. Some of the
later authorities have attempted a reform in the nomenclature of the
muscles, based on their supposed uses, and have only added to the previous confusion. Adductors and abductors have been so multiplied
that it would seem that a horse, like a crab, was made to go sidewise.

Anatomy will be treated no further than is necessary to demonstrate
locomotion; and those who would pursue it further, and those who
would be more minute in their knowledge of structure, must dissect
for themselves.

The writer has already had occasion to allude to *design*, and will
have frequent necessity for doing so in describing the complicated
mechanism by means of which locomotion in the higher orders of
animals is effected, and he wishes it understood that he uses that term
in its literal and highest signification. He does not shrink from the
use of terms that imply an intelligent Creator and all-pervading Spirit,
who, from the beginning, established the foundations of the earth, and
who, in incomprehensible wisdom and power, has fixed the laws which
govern the organic world from the beginning through all its changes.

In using the term " higher orders of animals," he follows custom. If
that distinction is founded on the complexity of his locomotive powers
or organization, then man could not justly claim the first rank; for if
his preservation had depended upon his speed in locomotion, he would,
in the long struggle for life through which he must have passed, have
taken his place in the earliest paleontological deposits.

It may seem presumptuous to compare objects, the lowest of which
are beyond our comprehension. The finite cannot comprehend the
infinite; there must be a limit, in the nature of things, to all inquiry
into the phenomena of life. If physical science could determine the
laws of that which is hyperphysical, then to its court we might

carry all cases involving ethical or æsthetical questions, and form
might be confounded with color. To this *pons asinorum* all the old
writers on animal mechanics came. They would test vital force by the
laws governing the motion of the pendulum or those of gravity. If
physical science could establish the laws and solve all the questions
that arise in the investigation of vital phenomena, and algebraic expres-
sions could represent the unknown quantities, the task would be easy.
We could calculate the force of the right arm of a warrior as we could
the weight of his sword; but when that arm descends, it falls with
more than the force of gravity. There is a power that must enter into
all our estimates of vital force, and that is the *will.* It cannot be
ignored in any calculation on animal motion ; and yet who can estimate
it, weigh it, and formulate it, as in the exact sciences?

Thomas Starr King used to tell a story of a countryman who
attracted the attention of a traveller by the fine physical development
he displayed, and of whom he inquired his weight. "Well, stran-
ger," said he, "ordinarily I weigh two hundred and thirty pounds, but
when I am mad I weigh a ton."

The progress that science has made in every department, and is
still making, is wonderful, and who can say where it will end? But in
the knowledge of the laws which govern the origin of life, the vital
organs and their functions, of the nature of that force by which one
form becomes altered or modified by the altered conditions of its life,
it has made no progress since the days of Job.

The whole question of life and vital force is still a great mystery,
although it is receiving at this time the concentrated attention of the
most intelligent naturalists of all nations. There are not many who
deny that organic forms may be modified within certain limits by arti-
ficial means. There are many who believe that all organic beings, of
whatever nature, had their origin in the most rudimentary element, as
a cell possessing certain inherent tendencies to develop by aggregation
into other and higher forms, unequally modified in various ways by
surrounding influences, with a tendency to variation by imperceptible
degrees in every direction, the useful variations favoring the existence
of the individuals possessing them. This idea has become familiar

under the terms "natural selection" and "survival of the fittest." This hypothesis does not presuppose design, and denies a Creator. Under the name of "Darwinism" it has become popular and invaded all ranks. It found the soil of Germany especially fitted for the propagation of a theory of such an atheistic character, and it was proposed at a meeting of the Society of Naturalists at Munich, a few years ago, to teach it in the national schools. It has become so generally diffused in our own scientific circles that a reference to a Supreme Being in an essay read before a society of naturalists would be considered to be a poetic license, if one had the courage to make it; and nature is usually personified to meet the necessity. We have long been familiar with the reference to the laws of nature, and we now begin to hear of the laws of evolution. In all ages there has been a tendency on the part of the masses to follow some leader whom they desired to do their thinking for them; to pin their faith to his, or what they supposed to be his: it is no less so in the scientific circles than in the religious. Dogmatism seems to be leaving the latter to attach itself to the former; at all events, it is inherent in the human mind; no person is utterly free from it; and to appeal to the opinions of those whom we believe to be better informed, rather than to examine the foundations of those opinions, has been the vice of all ages.

It is well known that faculties and functions are strengthened by use and weakened, or altogether lost, by disuse. We shall look in vain for proofs of an organ changed in the mechanical principle of its construction, or one evolved by imperceptible degrees where none existed before; but we shall, on the other hand, find proofs in anatomy that the changes could not have been gradual. Every stable-boy knows that qualities are transmitted by heredity, and that desirable ones may be bred by judicious crossing within certain limits; and he knows as much as any one of the force, or influence, by means of which this is brought about. Speculation should not be confounded with science, as was said by Virchow, or science will lose its claim to the respect of mankind; and this whole question of evolution is speculative when carried beyond proof; and science, when it crosses the vital boundary-line, is lost in speculation. We know that organic

3

matter is subject to physical laws like other matter: it is attracted
by the earth, and will fall with a force as great as if it were inanimate,
and is equally subject to the law of falling bodies; it acquires momentum, and its momentum is equal to the weight multiplied by
the velocity, the same as that of a railway-car, or a cannon-shot;
and when vitality leaves it, it is resolved to its original elements,
oxygen, carbon, etc., which the chemist can prove by analysis. But
has the most skilful chemist ever been able by synthesis to restore
the lost element, the vital spark? Has he ever been able to imitate
the products of that vital laboratory the stomach, and form the
aliment that replenishes the blood?

With all the knowledge of physics ever acquired by man, can he
make a pump so perfect as the heart, — that organ that forces the
blood loaded with fresh sustenance to every part of the body? And
what does he know of that power that has kept it in alternate action
and rest every instant since before the earliest memories of his child-
hood?

He has been familiar with the laws of optics for centuries, and has
made instruments of glass and metal, in imperfect imitation of the eye
of an animal, to exalt the powers of his own vision; but what would
not an optician give to be able to construct a concentric achromatic
lens, with automatic power to adapt itself to the distance of objects,
such as the eye of the lowest of the vertebrates?

Acoustics is another of the physical sciences of which man is a professor, and he has just invented an instrument by means of which he
can communicate in ordinary vocal sounds to a person miles distant.
He has recently invented another, by which he can register and preserve the intonations of his voice to be returned to him at will at any
future time; but that most wonderful instrument, the ear, he can only
wonder at and admire. Without it the world would be without music,
voice, or sound; the faculty of speech, our consciousness, memory,
imagination, affection — but it is needless to multiply this class of
facts. In nothing does man show himself to be the creation of an
intelligent power more than in his own creative faculty. How great
have been his achievements in mechanics! But what comparison does

the highest bear to the locomotive apparatus or machinery of the horse, with its compound system of levers, pulleys, tendons, springs, and muscular powers, and that marvellous ingenuity in arrangement to produce results which man has not been able to understand until now, and all set in motion through telegraphic communication distributed to every muscular fibre, and the whole of this complicated system of organs co-ordinated and controlled by one central will? Another incomprehensible mystery of life is, that this complicated machine should possess the power, not only to preserve and protect itself through a long life, but of reproducing from generation to generation indefinitely, and transmitting to posterity its own peculiarities of form and mental qualities!

Does the whole organic world furnish no proofs of intelligence and design, that we must be told that all these marvellous manifestations of both are but the inherent properties of matter,

"And that were true which nature never told"?

If it were an "attainment and an aim" to escape moral responsibility by getting rid of a creator, do we approach any nearer the solution of the question of the origin of life by removing it farther off into the mytho-geologic eras? Or is the difficulty in any way diminished by attributing to matter all the high intellectual functions that have been by unschooled people in all ages ascribed to supernatural powers?

Can the microscopist, when he discovers vibriones in a vegetable infusion, or protoplasm in a drop of serum, be excusable for running naked, like the philosopher of Syracuse, through the streets, shouting "Eureka"? Can one who finds a shingle or a brick claim that he has discovered the cause of a house? *Let him account for the origin of the brick and the shingle!*

Because a fossil skeleton of a four-toed horse, which failed to connect his species with our time, has been found in the fossiliferous deposits of the interior of this continent, does it follow that our noble soliped had an origin less remote and independent, or that he found it necessary and practicable to concentrate his four toes into one, or succumb to the altered conditions of his life?

All science, in whatever department of knowledge, is retarded much by the ignorance and zeal of the multitude who follow on the heels of genius. Medicine has its mountebanks, who are dragging a noble science into public contempt ; religion has its harlequins, and natural science its buffoons, who, as itinerant lecturers, perambulate the towns as representatives of learning they do not possess, and put forth as proved truth the wildest speculations of enthusiasts, and call them science. It is very common to hear of the origin of man from the ape, as if the relation were a scientific truth, when in fact it is only a speculation ; and all the evidence so far collected from fossil remains as early as the tertiary deposits gives no confirmation to the speculation. As far away as any trace of the prehistoric man has been found, he was as perfectly developed as he is to-day, and as far removed from the ape.

Darwin is not responsible for what is known as Darwinism. He is a model for a naturalist, collecting facts and placing them in their relation, drawing his conclusions cautiously, and candidly admitting the difficulty when a fact antagonizes the hypothesis he is framing. Not so with his zealous disciples, who rush to their desired conclusions over his facts, as the fanatical Christians of Alexandria did over the last vestal altar of Greek philosophy.

Organic life is the result either of chance or design ; there can be no middle ground.* If the latter, the question of how it was brought about will never be solved by man, nor is it important that it should be. It is sufficient that a Supreme Intelligent Will is the author and sustainer of all, — a beneficent Spirit, who

---

* Virchow, who will be recognized as one of the leaders in the new departure in science and the cell theory of development, says : —

" This much is evident. If I do not choose to accept a theory of creation, if I refuse to believe that there was a special Creator who took the clod of earth and breathed into it the breath of life, if I prefer to make for myself a verse after my own fashion, then I must make it in the sense of *generatio equivoca* (spontaneous generation). *Tertiam non datur.* No alternative remains when once we say, ' I do not accept creation, but I will have an explanation.' If that first thesis is laid down, you must go on to the second thesis, and say, · Ergo, I assume the *generatio equivoca.*' But of this we do not possess any actual proof. No one has ever seen a *generatio equivoca* really effected ; and whoever supposes he has is contradicted by the naturalist, and not merely by the theologian." — PROF. VIRCHOW, in a lecture delivered before the German Association of Naturalists and Physicians at Munich, 1877.

"Warms in the sun, refreshes in the breeze,
Glows in the stars, and blossoms in the trees,
Lives through all life, extends through all extent,
Spreads undivided, operates unspent " ;

who has endowed us with faculties to admire the beautiful, the good and true, to know *why* so many things are as we see them, but none to know *how*.*

Having given some of the reasons for his belief in the spiritual origin of the organic world, the writer claims his right, whenever he has occasion in the following pages to do so, to speak, without danger of being misunderstood, of design or contrivance in the same sense that he would when referring to similar manifestations of design in a humanly constructed machine.

In a theory of evolution, as the expression of the method in creation, the writer has little doubt that the thoughtful mind will in due time rest satisfied.

* " The consciousness of an inscrutable power, manifested to us through all phenomena, has been growing ever clearer, and must be eventually freed from its imperfections. The certainty that, on the one hand, such a power exists, while on the other hand its nature transcends intuition, and is beyond imagination, is the certainty towards which intelligence has from the first been progressing." — HERBERT SPENCER, *First Principles*, 2d edition, p. 108.

" When the remarkable way in which structure and functions simultaneously change is borne in mind, when those numerous instances in which nature has supplied similar wants by similar means are remembered, when, also, all the wonderful contrivances of orchids, of mimicry, and the strange complexity of certain instinctive actions, are considered, then the conviction forces itself on many minds that the organic world is the expression of an intelligence of some kind. . . . Organic nature then speaks clearly to many minds of the action of an intelligence resulting, on the whole and in the main, in order, harmony, and beauty, yet of an intelligence the ways of which are not as our ways." — ST. G. MIVART, F. R. S., in *Genesis of Species*, pp. 272, 273.

" There is something in organic progress which mere natural selection among spontaneous variations will not account for; this something is that organizing intelligence which guides the action of the inorganic forces, and forms structures which neither natural selection nor any other unintelligent agency could form." — MURPHY, *Habit and Intelligence*, Vol. I. p. 348.

CHAPTER II.

It is proposed to present as concise a view of the locomotive organs of the horse as may be consistent with a proper knowledge of the parts, and the functions they perform in progressive motion.

There can be no just appreciation of the qualities of a complicated machine without a comprehensive understanding of its construction, and the manner in which each of its parts acts to produce the compound movement for which it was designed. So, in order to understand the paces of the horse, we must understand the action of all the parts of the machinery by which they are produced. It need not be said that it is very complex, and has never been understood, for the reason that the motions themselves have been altogether misinterpreted.

This study of the mechanism of the horse is a necessity which will become apparent to any one who undertakes to analyze these movements by the aid of any manual of anatomy yet published. The distinction of muscles into adductors, abductors, extensors, and flexors gives a very inadequate idea, and sometimes a very erroneous one, of the action of the muscles to which those terms are applied, as well as to their general agency in locomotion. In fact, these terms are used to express the action abstractly with reference to the bones to which

they are attached, and not sufficient attention has been given to their
action in correlation to the others with which they are coworkers.
The forces employed in each limb, considered alone, are very complex.
The same muscle may be an extensor at one time and a flexor at another in the same stride, as we shall show further on.

In order to enable the reader to understand the muscles and their
relations without too great a tax on the powers of abstraction, the ser-
vices of Mr. William Hahn, a Düsseldorf artist, were secured to delineate
the most important muscles as they were exposed in dissection ; but no
skill can do justice to the nacreous tints of the tendinous envelopes
of the deep muscles.   With all the aid which art can render, the
complicated mechanism of the horse cannot be presented by written
description in such a manner as to dispense with a little close attention.
A perfect familiarity with the subject, so as to enable one to carry
the plan of the whole machine in the mind, can only be attained by the
aid of dissection.   A knowledge of the construction of the machine
is imperative upon one who would comprehend its action.   It is
as necessary as for an engineer to understand the construction of his
engine.   With that knowledge one can understand the elements of a
horse's strength and speed, analyze his movements, and appreciate the
source of the danger from injury in great trials of speed.

Let us first review certain physiological and general anatomical
facts, well, but not so generally known, as could be wished.   The mechanical parts divide themselves into two classes, the active and passive.   The passive parts are the bones and ligaments ; the active parts
are the muscles in which dwells all the power.

Of the bones it may be said, in general, that they are the levers on
which the muscles act, and by means of which their power is made
available ; their form depends upon the uses which they are designed
to serve.   When intended for bases of action, they are thin, angular,
and ribbed, like the shoulder-blade, or scapula.   When they are to serve
as columns of support, they are cylindrical ; and as there is always
the utmost economy used by the Creator where it is needed, they are
made hollow, for it was known, as long ago as the first mammal was
made, that there was no loss of strength as a support in being so con-

structed. It was long afterwards discovered by man, and the law was learned by him, that the lateral strengths of two cylindrical bones of equal weight and length, one being solid and the other hollow, are to each other as their diameters; and the spaces in the shafts of these bones, being needless for the purpose of support, are made depositories of fat or marrow for fuel, — literally, *coal-bunkers*, — as are all the angular spaces not needed for more important uses throughout the body, by means of which heat is developed, which primarily is the source of all motion in the animate as well as the inanimate world.

The extremities of these bony columns are spread out to give broader articulating surfaces; at the same time the single hollow of the shaft is divided into innumerable small ones, so that greater strength is attained to resist the wrenching force to which they are liable, without increase of weight; roughened ridges, spines, and protuberances * are formed to give greater surface for the attachment of muscles. For the purpose of still further increasing the surface for attachment of muscles, supplemental bones are added, as in the splint bones, — or, as they are called by anatomists, the *little metacarpels*, — which not only serve to widen the articulating surface, but, by a strong ligamentous membrane that connects them with the main pillar, give the necessary space for attachment of important muscles, and where the distance from the centre of motion renders the reduction of weight very important, as the rapidity with which these extremities move increases greatly with the distance from the centre of motion. The bones are composed of animal and earthy matter, in the proportion of about one of the former to two of the latter. If the proportion of the former is increased, they will bend under the force applied to

* Atheists maintain that *function makes the organ*; but how can we conceive of function without previous conception of the organ? What conception can be formed of sight without the existence of the eye? It is held by them that the roughened ridges and protuberances of bone are developed by traction of muscles upon the bony surfaces. If this is so, why is it that the surface of the bone above the acetabulum which receives the insertion of the *rectus femoris* is smooth? It certainly is not because there is want of traction on the part of that muscle. On the other hand, the *tensor vaginæ femoris* and the *superficial gluteus*, whose insertions are low down on the femur for the necessary leverage, must find room between other muscles, and a rough protuberance is formed to give the most surface for attachment in the least space.

them; and if the proportion of the latter is increased, they are liable to break. Variation from the normal proportions is the result of disease, and is more common in the human family than among quadrupeds.

The bones are covered with a compact, inelastic fibrous membrane, the *periosteum*, which adheres so closely to their surfaces that considerable force is required to detach it. This membrane serves not only to nourish the bones through its blood-vessels, or vascular system, but to strengthen them and increase their elasticity. The Californian Indian adopts the same method, for the same purpose, in the construction of his bow. In studying the architecture of the skeleton, as a whole, it will be found that no element of strength is wanting, or principle of mechanics violated, in its structure. The bones are arched or bent when such forms give greater strength. They are connected to each other by a strong tissue, so flexible as to allow of the greatest freedom of motion, but inextensible, and, under all ordinary use, too strong to be broken or detached from the bony levers whose motion it is designed to limit. It is, however, sometimes torn, either completely or partially, in dislocations or sprains; and the slightest injury to this tissue is a serious accident to an animal whose value depends on the soundness of his locomotive organs.

The extremities of bones which move upon each other, as at the joints or articular surfaces, are covered with a peculiar formation known as cartilage. It is insensible in a state of health, and very elastic to pressure; thickest where most exposed to concussion, and covered with a membrane which secretes a glairy fluid adapted to lubricate the opposing surfaces and reduce friction. These joints are all closed to the admission of atmospheric air and all foreign substances, for their admission would soon cause serious injury.

The joints are divided by anatomists into several classes, according to their mechanical construction. Some are simple hinges, admitting of motion in one direction only, as those of the lower parts of the extremities. The heads of all the four columns of support are provided with a kind of joint known to mechanics as the ball and socket. This form admits of the greatest freedom of motion in every direction; but the motion is limited in extent by capsular ligaments

4

which surround the joints as a continuous collar, whose borders are attached to each of the bones so far from the opposing surfaces as not to intervene, and yet not so far that they may not limit the motion to its needs. These capsular ligaments serve another useful purpose. Being air-tight, when the limb is off the ground it is supported in its place by the pressure of the atmosphere, — estimated by Borelli to be equal, in the hip joint of a man, to a lifting force of twenty-six pounds. The force thus gained is set free to be employed in locomotion. Each joint constitutes by itself an interesting subject for study, as they all differ in some important particular, according to their uses. The construction of the hock joint is quite unique, and has no analogue in man; and that of the hock of the ox is quite different from that of the horse. The interlocking grooves are oblique, so that when the posterior extremity is brought forward to pass its fellow, fixed upon the ground, it is carried obliquely outward, independently of volition; and when all danger of interference is passed, and the limb is again extended to reach the ground, the foot is carried obliquely inwards, to resume its place under the centre of gravity. This will be referred to more fully when considering the action of the posterior extremity. The construction of the joints at large would serve as a subject for a monograph of great interest; but to be fully understood it must be studied *cusis in manu.*

A detailed description of the bones will not be attempted. They are proverbially a dry subject; but for the convenience of those who require it, a reference plate is presented, lithographed from a photograph; and it is hoped that it will, through the eye, give the necessary information to enable the reader to understand the mechanical movements without the study which abstract description would require. But the vertebra, or spinal column, as the keel or bed-plate connecting the various parts of the machinery, requires further attention.

The term "spinal column," as applied to the skeleton of quadrupeds, is a misnomer, derived, like most anatomical names in comparative anatomy, from its analogue in man. The spine being horizontal in quadrupeds, and not vertical, as in man, the term "column"

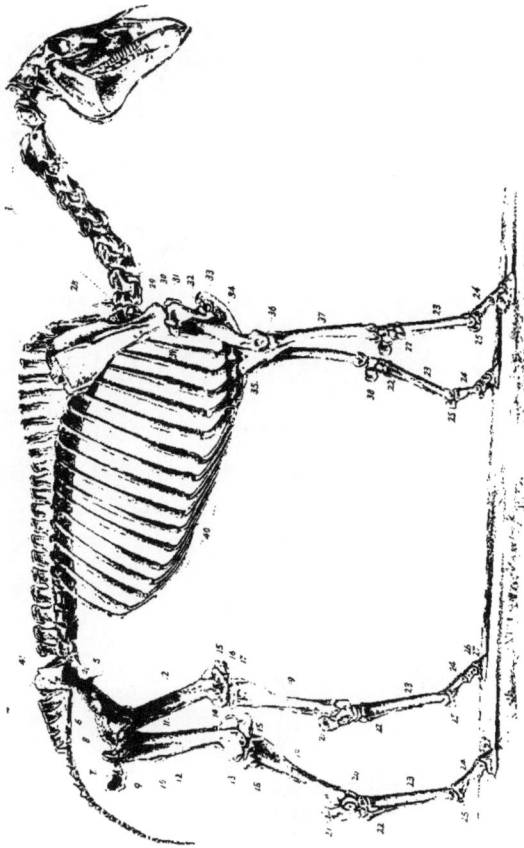

is inapplicable to them. The word "spine" is also objectionable, as it is derived from the processes which superficially mark its course. There seems to be no objection to the term "vertebra," as a collective noun applied to the whole or any number of its parts. As it is the keel and connection of the various parts of the animated engine, so it is the term from which has been derived the name for the whole division of animals to which quadrupeds belong, — Vertebrates.

The vertebra of the horse is divided into five groups, differing materially in their mechanical, even more than in their physiological relations. These groups are the Cervical, the Dorsal, the Lumbar, the Sacral, and, lastly, the Caudal. The cervical vertebræ have an important relation to locomotion, second to no other division. They are provided with spines along the median plane, as are all the vertebræ, and transverse projections or processes, which afford attachments to ligaments to maintain their relative positions; and with important muscles, as will be shown in a subsequent chapter. There is great freedom of motion of these bones upon each other, in comparison with those of the next two divisions, especially at the articulation with the head and the first vertebra of the trunk. This last is a ball and socket joint of a peculiar construction, to enable the animal to reach the ground, as in grazing and drinking.

The second group is the dorsal, and it consists of those vertebræ that are articulated with the ribs. Like the cervical, these are provided with transverse processes, which serve not only for muscular and ligamentous attachments, but as braces to the ribs. The spinous processes are longer than those of any of the other vertebræ, especially along the withers, where the suspending muscles of the anterior extremity originate. It will be apparent to the most superficial observer that the motion, either lateral or vertical, of the dorsal vertebræ upon each other is very circumscribed, being limited in a vertical direction by the long spinous processes and their intermediate inelastic ligaments, and in a lateral direction by the transverse processes and their articulating ribs. The next division is that of the lumbar region, or the vertebræ of the loin, with which there are no connecting ribs. As the former group was more intimately related to the thorax, so those of the lum-

bar are in the same relation with the abdomen. Their broad and
long transverse processes afford a protecting roof to the abdominal
viscera, and give attachment to important muscles of locomotion on
the under surface. There is very little movement of these bones
upon each other, even less than in the dorsal series, — so little that
bony union takes place between them in old age; and the elastic
cartilages that, at an earlier period of life, were interposed between
each of the vertebra become degenerated into bony matter, and that
condition obtains technically known as *ankylosis*.

The next series, and fourth in order, is the Sacral. Though in
the embryotic stage the sacrum is developed from several centres as distinct vertebra, yet before birth they are united into one broad triangular
bone, which, uniting with the iliac bones on each side, and the pubic
bones in front, forms the ring known as the pelvis. It is in the lower
or pubic portion of this pelvis that the cuplike cavities are formed
into which the heads of the hip bones are lodged, and where the force
of the levers of the posterior extremities is applied. The difficulty
in locomotion that would be experienced from the want of flexibility
of the spine, especially in old age, is obviated by the freedom of
motion that is secured in the articulation of the last of the dorsal
vertebra with the sacrum. This is what is known as the "coupling,"
as it unites the two distinct systems of locomotive organs, the anterior
and posterior extremities. In the skeleton the connection seems very
slight ; but the ligamentous connections are very strong, and the long
muscle of the back (*longissimus dorsi* or *ilio spinalis*), reaching out
from its spinal attachments, lays hold of the hip bone (crest of the
ilium) on each side as far as possible from the centre of motion at the
coupling, the more effectually to limit the flexion at that point.

The last group of vertebral bones is known to anatomists as the
coccyx, from its resemblance in man to the beak of the cuckoo ; but as
the resemblance totally fails in the Mammalia and all other vertebrates,
we shall call them by the more general name of Caudal bones. They
have no function in locomotion ; but " thereby hangeth a tale."

Between all the vertebral bones is interposed a layer of elastic cartilage, of the same nature as that which covers the opposing surfaces

THE HORSE IN MOTION.

of the joints in the extremities. These cartilages by their elasticity admit of slight flexion of the vertebra, and they also deaden the force of the shock transmitted from the powerful impulses of the posterior limbs. As has been already stated, the flexion is limited by the ligaments which bind them to each other. This restriction of motion is necessary for the protection of the vital organs of the thorax and abdomen, as well as the great nerve trunk transmitted through a continuous canal above the bodies of the vertebra, and which is distributed thence to all parts of the body.

While the three central divisions of the vertebra may be curved slightly, they cannot be shortened, even temporarily, as may be readily seen; and the apparent shortening that takes place when the animal's limbs are gathered under him is an illusion. The elasticity of the cartilages and ligaments is greatest in the young; as age advances, these tissues become stronger and less flexible, and resist the movements of the joints; they are said to become "stiff." Hence the importance of early training to give greater sweep and freedom of motion. This physiological principle is made the basis of gymnastic training by acrobats, being commenced at a very early age; and the same is not lost sight of in the exercises of colts.

In contemplating the passive parts of the animated machine abstractly, we see the *results* of organic life; they are without sensibility or power of spontaneous motion; we are familiar with the mechanical principles involved in their action, and are impressed by the perfect adaptation of means to ends; we look upon them as we look upon the piston, connecting-rod, and crank of a steam-engine: but upon the muscles we look with far different thoughts; their action has no similitude in the inanimate world.

The general appearance of muscle is too familiar to every one to need description; its special vital property is *contractility*. The muscles are both voluntary and involuntary, but it is only the former that are concerned in locomotion.

If we remove a fragment of muscle from an animal recently killed and examine it closely, we shall find it to be made up of longitudinal fibres of a red color bound together by gray fibres of a different tissue.

If we lay this flake of muscle upon a plate and scrape it gently in the
direction of its fibres with a dull knife, we shall find upon the edge of
the knife a red pulp without apparent fibre or tenacity, and there will
be left behind a bundle of strong cellular tissue. It is to the former
that the tractile property belongs; the latter has no more active power
than other cellular tissue; yet this pulpy bundle of fibres, as muscle,
contracts under the stimulus of the will with almost inconceivable
power. Borelli estimated that the force exerted by the deltoid muscle
of man in supporting a weight held horizontally in the hand was two
hundred and nine times greater than the weight. Therefore a weight
of sixty pounds held horizontally requires an expenditure of contractile
force of the extensor muscles at the shoulder of more than six tons.
He demonstrated that the force of the extensors employed by a porter
in carrying a weight of one hundred and fifty pounds upon the shoulder exceeds three tons.*   It follows that this enormous power is
exerted on the extensors of each leg alternately.

The natural stimulant to the muscle is the will transmitted through
the nerves; but the will is not necessary to muscular contraction, as it
has no influence on the muscles of animal life or the vegetative functions of animals, and any of the voluntary muscles may be cut off from
communication with the brain by severing its nervous connection; yet
contraction may be excited in the muscle so cut off, and this may be
continued indefinitely by further division to a microscopic degree; still
the fibres will be observed to contract upon the slightest touch, so
closely are the nervous fibres interwoven with those of the muscle.

Electricity when passed through the muscle in a broken current is
a strong excitant to muscular contraction, overmastering the will, and
will even cause contraction after life has left it; but if the current is
continuous, it has no such power.

The muscular fibres are paralyzed by certain poisons, and stimulated
to violent contraction by others; and in disease, as tetanus, they may
be so violently stimulated as to be torn asunder. This subject, though
very interesting, is leading away from the special inquiry to which we

* Cyclopædia of Anatomy and Physiology, art. "Animal Motion."

are limited. The muscles are subject to fatigue, and are unable to respond indefinitely with equal force to the will.

Muscular fibre has other properties to be considered in relation to motion. Its contractility is limited to one fourth * or one third † of the length of the fibre, and with a power proportioned to the area of the transverse section of the muscle. It will be found that the relation of length to thickness is as action to power.

Deep-seated muscles are often attached to the bones upon which they act directly; but as there is insufficient space on the surface of the bones for all that depend upon them, the extremities of the muscles are often changed into tendon, — a substance altogether different in its mechanical properties, being compact, very flexible, and incapable of elongation, in order that it may not give away the contraction effected by the muscular tissue. By means of this tendinous tissue the power of the muscle is transmitted when necessary to a considerable distance, or its direction may be changed by the tendon passing through a sheath or groove, as a pulley, over an angle. In a humanly contrived machine it has been found necessary, when the direction of the action of the power requires to be changed, to use a friction roller or pulley; but nature has done better, and contrived a way to avoid friction and wear that human ingenuity cannot hope to rival. By these means the power generated in the heavy muscles is exerted at the extremities of the limbs where all needless weight requires such great expenditure of power to give it the needful velocity. The power which is conserved in the body as momentum would be lost in the extremities, for the motion of the limbs is arrested at every stride. ‡ The attachment of these tendons to the bones and the periosteum enveloping them is so great that detachment by natural means is not mentioned in works on farriery as among the possible accidents to which the horse is liable.

---

* Bishop.

† Bowman, Cyc. Anat. and Phys.

‡ If a weight of 25 lbs., sustained by the hand of an arm extended horizontally, requires the expenditure of an energy equal to 200 times that weight, or 5,225 lbs., what amount of muscular force is expended by the muscles of one of the extremities of a horse to move a 4-ounce shoe on his foot when he is trotting at the rate of a mile in 2 min. 20 sec.?

During the life of the animal the tenacity of the muscle is greater
than that of its tendon, but when vitality no longer animates it it may
be easily torn.

While the articular ligaments are subject to extension and elongation by early use and frequent tension, so that greater freedom of
motion than is normal is acquired, it is otherwise with the muscular
tissue and its tendons. By exercise within certain limits, at regular
intervals, and with proper nutrition, the thickness and power of the
muscles may be increased, and by neglect of these conditions they will
become thin and pale, while contraction will be feeble and not well
sustained; but they will not become elongated under whatever violent
and long-sustained exercise; they may increase in thickness, but not
in length. But for this exception to the rule the whole plan on which
animal mechanics was founded would have fallen to the ground with
the animal himself. Were the muscles to become lengthened by use
without corresponding increase in length of levers, the tension necessary to prompt action would be lost, and the effect would be similar
to that upon the tiller ropes of a ship were they to become relaxed.
What would be the effect upon the length of the bones in the period
of time contemplated by some it is useless to inquire, but we know
that the increase of muscular power by increase in the bulk of the
muscle takes place in a short period, and in the lifetime of the individual. But while the muscles and their levers will retain their normal relation of length during life in a healthy subject, that balance is
sometimes lost as the result of injury. A child has been run over by
a wagon; the wheel has passed over the muscles of the calf so as to
disorganize the muscular tissue; in due time the injured part is restored to health, but the muscle does not develop fully; it is shortened,
and a form of club foot is the result, in which the person cannot, while
walking, reach the ground with the heel. The child has grown to
manhood, but no amount of use and no length of time will elongate
the muscle. Nature cannot elongate that muscle without anarchy.
The Creator works by law, and to claim an exception is virtually an
admission that we do not understand the law. But what He cannot
do without anarchy his creature can; he slips a tenotomy knife

beneath the tendon, severs it with scarcely a visible external wound, the muscular fibres retract the severed ends, and that ever-present, inscrutable power fills in the space left by the parted extremities of the tendon with new tendon, the organ is restored to proper length, and the deformity is removed. If, on the other hand, one of the bony columns of support be broken, — for example, the thigh, — the creative power called nature soon sets at work to repair the damage. A sequestrum, or casing, is formed around the broken extremities, consisting of inelastic bony matter, to fix them in their position as a temporary expedient, while the slower processes of the more thorough organization of perfect bone is effected, and the fracture is repaired, after which the sequestrum is absorbed and carried off through the circulation. While this change has been taking place in the bone, it would, without surgical interference, in most cases be shortened by overlapping through the contraction of the muscles on all sides of it. The consequence would be that the same disaster would be encountered as in the last case, where the muscles were supposed to be elongated from use; but another law is observed. The muscle that could not elongate will shorten, and the proportion between the length of the lever and the muscles which act upon it is restored.

It is said by Professor Marcy that "the comparison between ordinary machines and animated motive powers will not have been made in vain if it has shown that strict relations exist between the form of the organs and the character of their functions; that this correspondence is regulated by the ordinary laws of mechanics; so that when we see the muscular and bony structure of an animal we may deduce from their form all the characters and functions they possess." This statement, which *in the main* appears to be true, requires qualification. The form of many muscles is made to conform to the situation and relation of surrounding organs. Nature, while prodigal where she can afford to be, is economical where there is need of it. This is shown in numerous ways, and especially in the form and arrangement of muscles.

Beauty of form is never lost sight of in the construction of the horse; and even great sacrifices of mechanical power are made to maintain graceful lines, and that general contour of form that gave

to him his matchless beauty, — beauty so great that to the eye of a
superficial observer it is difficult to decide whether it is subordinate to
strength or conversely.   Both are developed in a perfect horse to such
a degree that he has been a favorite theme of poets and painters since
æsthetic culture has had a place in the history of our race.

Numerous instances might be referred to where use has been sacrificed to economy of space and to beauty; but they cannot fail to occur
to the mind of the anatomist; and it is premature to introduce them in
this place for the general reader.

CHAPTER III.

FROM the general observations of the last chapter we will proceed
to a consideration of the special anatomy, and analyze the locomotive organs of the horse; without this preparatory study it will be
impossible for any one to analyze its movements.

Those who have studied and suppose they understand this action
must study again. Let no one be turned from this subject by technical terms; they are indispensable in order to make one's self understood by those who have already made a study of anatomy, as well
as to those who would follow the movements by which the various
paces are performed, and speak of a horse in more intelligible terms
than the slang of jockeys and the stables. I think I am warranted
in the belief that we are on a new era in the history of our old
friend and fellow-traveller; the increasing interest that is felt in
America as well as Europe, and the impulse that is sure to be communicated by the wonderful revelations of the camera, justify me
in that opinion. I shall not follow the usual order of descriptive
anatomists.

It has already been stated that it is not the purpose of this essay
to teach anatomy any further than is necessary to demonstrate the

mechanism of the locomotive organs, and the manner in which the muscles act upon their bony levers to produce the movements in progressive motion.

The long muscle of the back holds the same relation to the locomotive muscles that the vertebra does to the bones; it is a very complex muscle or system of muscles; it is called by Chauveau the *ilio spinalis*, so named from its attachments. It fills the angular space on each side of the spinous processes, giving roundness to the back. It is very broad and thick over the loins, and is attached to the whole anterior border of the ilium and strongly to its crest, or the hip bone, as seen in Plate III., *q;* it is attached anteriorly to all the spines of the vertebra, as far as the neck, and a strong membrane, tendon-like in its construction, that is firmly fastened to the same bones. This tendinous membrane, called *aponeurosis*, has not been mentioned thus far, but it is tissue very important in its relation to the muscles; it differs from fascia in several respects, but specially in thickness and strength. It covers nearly all the superficial muscles, and its strength is so great that the muscular fibres are attached to its inner face as to a bone, and it serves them often the same purpose as fixed attachments.*

If one takes an elevated seat with the driver on a coach, and looks down upon the wheel horse nearest him, he can see the action of this muscle, and to the best advantage if the horse is trotting. It will be noticed that the spine is flexed in a serpentine manner as the diagonal legs move alternately. This movement is caused by the impulses given to the pelvis by the heads of the femurs alternately, which would seriously strain the articulation of the pelvis with the lumbar vertebra called the coupling, but for the action of the ilio spinalis, which contracts simultaneously with the impulse communicated to the opposite side of the pelvis, acting as a brace checking the wrenching violence of the action and preventing injury to the coupling. This is the function of the iliac wings, as

* A familiar example may be seen in a porter-house steak of beef. The part known as the tenderloin is a section of the psoas; that above is a section of the ilio spinalis overlaid by its aponeurosis.

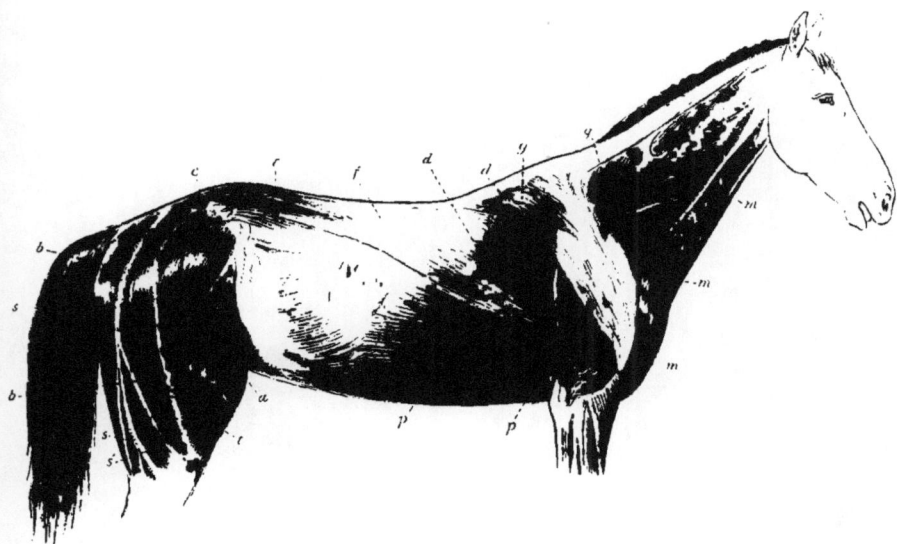

.

referred to in Plate V., *a, a, a*. The great mass of the muscle which fills the angular spaces on each side the spines is called into action in rearing, or supporting, the anterior half of the body when not supported by one of the fore legs. The greater part of the ilio spinalis is concealed in the plate by the great gluteus, *c, c, c*. The centres of motion between the vertebra are in the bodies of those bones which are most distant from the spines, and which form the rounded ridge of the backbone as seen in the great cavity of the trunk. In man they constitute the supporting column. The ilio spinalis muscle lies wholly above this axis, and its action abstractly would curve it downward; it can have no influence, therefore, in aiding to support a back load. The mechanical action of this long and powerful muscle is therefore, first, when they both act in unison to support the anterior half of the body while the pelvis is fixed by other muscles; and in the second place, when they act alternately, to counteract the wrenching effect of the propulsion of the heads of the thigh bones.

Before we proceed any further with the consideration of the muscles of locomotion, we must agree upon the signification of terms necessary to be employed.

The words "flexor" and "extensor" may be proper enough in some of their applications and express fully the action, but not in all. Some muscles act as flexors and extensors at the same time; others are extensors at one part of the stride and flexors at another; and some of the most powerful propellers in the whole machine are flexors, as we shall show in the course of this treatise. It will be seen that the actions of the muscular powers are sometimes quite too complicated to be expressed in one word.

The term "extensor" is commonly applied to all muscles whose action is to enlarge the angles and by so doing elongate the limbs; but this extension may be forward when the foot is in the air, or backward when the foot is on the ground. There is no word in use by anatomists to express the fundamental idea, propulsion. The terms "flexion" and "extension" will be used in the following pages to express the action of a muscle upon its attachments, without reference to its functions in locomotion. The words "adductor" and "abductor," meaning the function of drawing to or away from the vertical plane passing through the axis of the body, are well enough, but we must not be misled by the application of these names to muscles which may have such action to the extent only of five per cent of their work, and the rest, or eighty-five per cent, devoted to propulsion.

I have already referred to the misnomers in muscles ; they mislead the mind no less with regard to their action than to their form and construction. What can be more inappropriate than the names *semi-membranosus* and *semi-tendinosus*, meaning *half membrane* and *half tendon*, when applied to the muscles so named in the horse ? They are well enough when applied to the corresponding muscles in man, but in the horse they are not at all membranous or tendinous.

We should be glad to dispense with names altogether, and apply abstract or algebraic terms to avoid misconceptions, if practicable, but we must use such as are given, and, where there are synonymes, use such as are least liable to the objection referred to.

There is a group of muscles whose action is to advance the whole posterior extremity after the act of propulsion is complete. They are all deep-seated, with two exceptions.

The *psoas magnus* (Plates VI., VII., *a, a*) has its origin in the abdomen, along the under surface of the lumbar vertebra ; its fibres, which determine the course of its action, are directed backward and downward, and it terminates in a long tendon which is inserted into a rough ridge on the inner side of the femur, or thigh bone, just below the head of the bone ; another of this group is the *iliacus* (*c*, Plate VII.), which arises from the lower face of the ilium, or hip. The course of its fibres is similar to that of those of the psoas, but its origin being farther from the median plane, its direction is more inward to join the last-named muscle at the same point on the inner face of the femur. These two muscles are of delicate organization, and, though differing in form, unite in their function of flexing the femur upon the pelvis, and so carrying the whole leg forward. The *iliacus*, having its course more inward than the other, has the effect of carrying the free end of the femur outward, — the "stifle action," so important in the trotting horse.

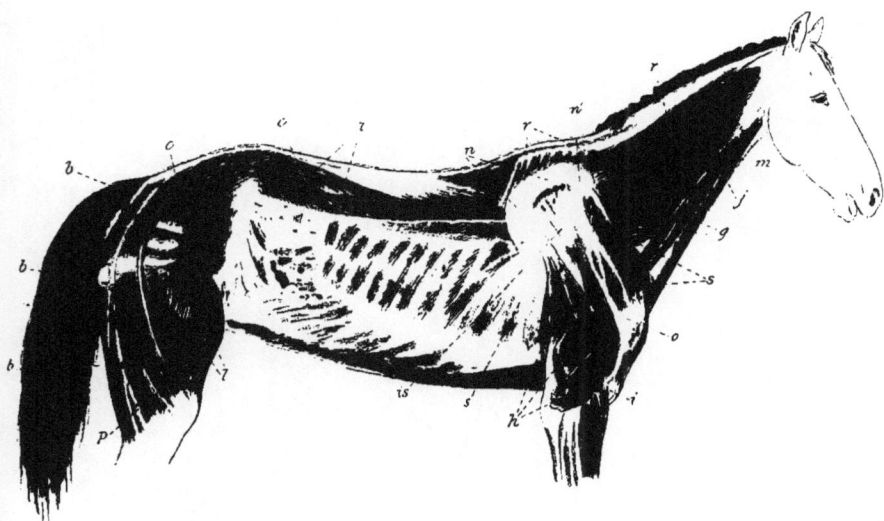

The *tensor vaginæ femoris* (Plate III., *a*, *a*) has its fibres spread out beneath the skin and the broad fascia of the thigh. It has its fixed insertion in the crest of the ilium, or hip; its fibres are about eight inches in length, and its weight not less than two pounds; its action, direct and indirect, is upon the thigh to flex that bone upon the pelvis; from the shortness of its fibres its action as a flexor cannot extend beyond three inches, but, being exerted at the commencement of the flexion, when its aid is most required, it is very useful. It is intimately associated locally and functionally with the superficial gluteus, which has one of its attachments at the hip bone, and another at the thigh bone, or femur, about one third of the distance from its head. This portion, therefore, acts with the last mentioned in flexing the thigh; the other branch extends alongside of the long vastus, filling the angular space made by that muscle where it crosses the great gluteus. (This is made clear by Plate IV., where the muscle under consideration is dissected away, along with the tensor vaginæ femoris.) It will be seen that it arises from the spine, in front of the origin of the long vastus, *v*, *v*, *v*, and its tendinous insertion is at *b*, or third trochanter of the femur (see skeleton, Plate II., *b*); the action of this division is therefore that of an extensor, and directly over the head of the femur at *c*, as we shall see when we come to consider the action of the posterior extremity as a unit in locomotion. The action of this muscle has been a controverted question. Blain teaches that it is a flexor of the thigh, Bourgelot classes it with the extensors, and Chauveau is of the opinion that it is an adductor. This confusion has evidently arisen from confounding the action of its two branches. From these two fixed insertions, so remote from each other, the fibres converge to the movable insertion at the ridge on the femur, as already stated, about one third down the length of the shaft, and between them the fibres of the muscle are lost in the fibres of the underlying muscle and barely distinguishable in the plate. The *form* of this muscle has never indicated its use in locomotion, but when removed, as in Plate IV., its value as an element of beauty is made apparent.

The *sartorius* (Plate VI., *b*) of the old authors, so called from its analogue in man, and so called in man because it is the muscle which enables him to assume the cross-legged position of a tailor, is named by Chauveau the *long adductor*. It has its origin on the tendons of the psoas muscles at a distance from the mesian plane equal, in the normal position of the animal, to that of its insertion at the inner head of the tibia. The distance of its corresponding origin in man would carry it fully seven inches farther outward and across the great body of the iliacus muscle. Its action, therefore, is simply as a flexor of the thigh upon the pelvis, but from its great length, eighteen inches, it has a sustained action in carrying the limb forward to a new position.*

There are some other small muscles, such as the *pectineus*, *small adductor*, etc., whose weight is so inconsiderable, and whose action is so near the centre of motion, that they cannot be supposed to have any special influence in locomotion. They are of more interest to comparative anatomists, but mechanically they are of small weight. The action, such as it is, seems to be allied to the last, or that of adduction, to preserve the balance between the adduction and abduction of the great propelling muscles, for it appears to be true that nothing in the animal economy was made in vain, and no vacuum exists. When the ancients propounded the law that "Nature abhors a vacuum," they "builded better than they knew."

In the complicated mass of muscular forces involved in each of the propelling limbs of the horse, it is impossible to determine whether adduction or abduction predominates: under the exercise of the will, either may do so; but when the mind of the quadruped is directed to some exterior object, to the attainment of which the co-ordination of all the locomotive forces are necessary, the adductor and abductor action of the muscles may be considered literally *side issues*, and the

---

* Herein lies a curious conundrum for the Darwinians of the atheistic school. If changes were by insensible degrees, how did the origin of this muscle become transported from the superior spinous process of the ilium in man, to the tendons of the psoas muscles across that body of the defenceless iliacus? That it should have been effected by imperceptible degrees seems entirely out of the question; and as there is a doubt as to priority in order of descent, or ascent as the case may be, we will take the liberal side, and admit that the two families, Equus and Homo, are of equal age and still evolving, but like parallel lines they can never meet; that the Equus can never be so much as a ninth part of a Homo, or a Homo so much as an Equus asinus without tangling his legs worse than with a too free use of his favorite beverage, or an interchange of the origins of the sartorius.

propelling forces alone are called into play, and every muscle "of the line" has to contribute its part, and the action is automatic.

The muscles not employed as propellers or carrying weight are few and small, as we have seen, bearing no comparison to the others. We will consider the latter in their order, commencing with the *great gluteus* (Plates III., IV., V., *c, c, c,*). It is a muscle of the first rank. As seen in the plates, it reaches forward over the loins and adheres to the strong aponeurosis, or tendinous membrane overspreading the ilio spinalis. It passes over the concave border of the ilium, or ridge between the hip and the angle of the croup, covers the upper surface of the ilium and the ligaments that cover its openings, and is at-tached to the spines of the lumbar vertebra and those of the sacrum; it is also attached to the strong aponeurosis that covers it externally like all superficial muscles of the back. This aponeurosis is repre-sented as dissected away in the plate. Its fibres all converge outward, downward, and backward to their insertion into the great trochanter behind the head of the bone, best represented in Plate V., *c, c, c.* (The great trochanter is so largely developed that it forms the short arm of a lever, bent at almost a right angle to the shaft of the bone, whose length does not exceed four inches from the head of the bone as a fulcrum.) The length of its longest fibres is twenty-six inches, and its average weight in two well-bred mares * was found to be sixteen pounds. It occupies a very advantageous position to give speed to the movements of the leg. The length and volume of its muscular fibres enable it to keep up a sustained action from the time the hind foot takes the ground under or in advance of the centre of gravity, until it leaves it after completing its propulsive effect. When the foot is off the ground it furnishes the sinews of war, offensive and defensive. The distance from the insertion to the fulcrum or head of the bone being so short, it causes the foot when free from the ground to move with great velocity.

---

* The weight of the animal from which the measurements and weights of muscles are given was about 1,100 lbs. These figures must not be considered absolutely correct, but relatively. In a horse regularly worked the muscles will be found to be heavier than in the better bred but idle ones sacrificed on the altar of Science.

6

If the reader will refer to Plate III. he will see only a portion of
this muscle; its extent forward is concealed by the pearly-colored aponeurosis which completely covered it and is only partially dissected
away; and by comparison with Plate VIII. and the skeleton, Plate II.,
he will find little difficulty in understanding the relations of this muscle
with the surrounding parts. In the succeeding Plate IV. the whole
outer face of it is exposed except the extreme posterior border, which
is covered by the long vastus muscle crossing its fibres diagonally; the
concavity in the ridge of the ilium from b to g, Plate VIII., shows also
the aponeurosis which covers the ilio spinalis and which serves as
a base for the attachment of the gluteus forward of the ilium. At
Plate V., h, are seen the attachments by tendon of the great gluteus to
the trochanter. (See skeleton, Plate II.) The centre of motion, or
head of the femur, for the posterior limb is a little in front of this, lies
deeper, and cannot be felt externally. This trochanter, therefore, is
relied upon by horsemen as a point for measurement, and is known to
them as the " whirlbone." Referring again to Plate VIII., the severed
tendons of the great gluteus may be seen at c, c.

The *deep gluteus* is well shown at P, Plate VIII. It arises on
the shaft of the ilium, and its fibres follow the course of that bone and
adhere to it as they descend. Its muscular fibres are intermingled
with tendinous bands following the same course, and the insertion of
the muscle is into the neck of the femur, or thigh bone, just outside
of the capsular ligament. Its curious construction of mingled bands
of tendon and muscle gives it the properties of both, the passive resistance of the former and the active aggressive force of muscular fibre.
The spiral course of its fibres indicates that it is intended to rotate the
leg outward, but more especially to hold the head of the femur in its
socket. Its influence in locomotion must be small.

The *long vastus* is second only to the great gluteus in weight,
its equal in length, and from its great advantage of position much
superior to it in effective power to perform the work required of
it. Its position may be seen in v, Plates III., IV., IX., and in Plate
V. its absence is more conspicuous than its presence could be. Its
insertion is into the external condyle of the femur (see Plates II.

and V.), and its relations are so perfectly shown in the plates as to scarcely require description. Plate IV. shows the superficial gluteus removed and the anterior margin of the long vastus exposed. It has its origin on spines of the sacrum posterior to those occupied by the superficial gluteus; it fills the deep fossa anterior to the tuberosity of the ischium, and overlaps the hip joint four inches; being lodged in this deep fossa, its position is fixed at that point; its direction is then changed so as to run downward and forward until it reaches the lower end of the femur, where its tendon is confounded with that of the patella.

As thus described, the posterior branch, which is admitted by anatomists to be distinct in structure and function, is detached. (It is marked s' in Plate V.) This is done for reasons which will be given when we come to consider the semi-tendinosus. Its weight, as so limited, is nine pounds, its length twenty-six inches. The space occupied on the surface in front of the tuberosity of the ischium is eight inches, or four inches over the trochanter of the femur, and the circumference of the body of the muscle at that point is fifteen inches.

It is nearly uniform in thickness throughout, except as its muscular fibres give way to tendinous ones toward its lower insertion. While the great gluteus has some of its fibres measuring as long, the great mass of them, on which its strength depends, are not half that length. The concentration of the fibres of the gluteus before their insertion into the trochanter is very great, and as their power depends upon their number and not upon their length, that of this muscle is enormous. Though it acts on the short end of the lever, the line of its action is very direct. But the vastus acts upon the extremity of the long end of the lever, and from the great length of its fibres sustains its action for a long time. These muscles hold a very interesting relation; they supplement one another. The power of the gluteus is effective in giving velocity, as in kicking; that of the vastus is effective in pushing the body over the foot on the corresponding side, when it is fixed upon the ground, as in rearing and leaping; in the hare, whose mode of progression is by a suc-

cession of bounds, it is developed enormously in comparison with
the gluteus.

The *semi-tendinosus* is represented in Plates III., V., IX., s, s, s,
where its situation is shown immediately behind the vastus. It
has two origins, one from the sacral spines and the first of the tail
bones or their ligaments, the other from the lower face of the
ischium (Plate V., i), below which they unite. It divides into three
branches; the central is attached to the strong fascia covering the
muscles of the calf, the other two reach forward to be attached to
the same common fascia, one on the inner and the other on the
outer face of the leg; the latter is spread out as far forward as the
insertion of the long vastus; the inner to a corresponding position
on the inner face. These lateral branches overlay the muscles of
the calf, or gastrocnemii, and give that compressed form that distin-
guishes the calf of the horse's leg. It is a powerful muscle. Its
weight is eleven pounds. The distance of its origin at the spine
to its insertion at the head of the tibia is twenty-eight inches.
The part of the muscle which has its origin at the ischium, to the
same point of insertion, is nineteen inches, and its greatest circumference is ten inches.

The action of this muscle cannot be represented by any abstract
terms. It has two functions: it lifts the leg when the act of
propulsion is complete, flexing the leg upon the thigh until the
line perpendicular from the centre of motion is passed, when it
relaxes, while the extensor proper of the leg, the *triceps femoris* (t, t),
carries the foot to a new position in advance. As soon as the foot
is upon the ground and the limb feels the weight thrown upon it,
then the full power of this muscle is called into play, no longer as
a flexor, and not as an extensor, nor even as a propeller, but as a
supporter, which character it performs until the direction of its fibres
passes the perpendicular, when they cease to act until the next stride
begins, so that when the foot is off the ground in the first quarter
of the stride it is a flexor; it is inactive in the second quarter,
and a supporter in the third, while it plays no part in the fourth.
The importance of the proper understanding of the action of this,

as of other muscles of the haunch, will be appreciated when we come
to the consideration of the fast paces.

The external branch of the semi-tendinosus has by all anatomists
been claimed as the posterior part of the vastus, while it was admitted to be anatomically and functionally distinct.   There is really no
relation between them except in their juxtaposition and in their superficial appearance.   Their connection is by a thin layer of cellular
tissue, while the connection between the branch in question and the
semi-tendinosus is most intimate, the partition being an aponeurosis
to which both are attached, as in penniform muscles, from which it
is impossible to separate the muscular fibres without laceration.   I
have no doubt that the point will be conceded by all anatomists when
their attention is called to it, especially since it is shown that the
annexation I propose makes a complete organ of the semi-tendinosus,
with all its parts acting in perfect accord.

The *semi-membranosus* adjoins the last-described muscle and is
concealed by it in Plate V.   Their relation is seen in the posterior
view, Plate IX., *t*.   This muscle also has two origins like the last,
but that at the spine is by a thin tendon, and this branch is small
(Plate VI., *a*).   The great mass of the muscle (*t*, Plate VII.) arises
from the lower surface of the ischium (Plate VII., *c*).   It is thin pos-
teriorly where it overlaps the semi-tendinosus (at *t*, Plate IX.), but becomes thick where it unites with the so-called great adductor (*g*, Plate
VII.).   The lower insertion is broad, the posterior portion of it is into
the fascia of the leg, and the anterior by tendon along with that of the
great adductor into the interior condyle of the femur opposite that
of the vastus; its weight is six pounds.   The thin posterior portion
of the muscle acting on the fascia of the leg flexes it like the semi-
tendinosus, but the great mass of it acts in unison with the great
adductor (*g*, Plate VII.), with which it is so closely united that it is
difficult to separate them.

The *great adductor* also rises from the ischium in front of the
last described, and is inserted into the internal condyle of the femur;
its weight is three and a half pounds, and its fibres are fifteen inches
in length, though fibres are thrown off along its course to the femur,

on which it acts as an adductor when the animal is at his ease, but the
joint action of these two muscles is as supporters. They have no
attachments forward of the centre of motion at the head of the femur,
but like the semi-tendinosus they permit the limb to be advanced to
the extended position to support the centre of gravity, and then, in
common with all the great muscles of the posterior extremity, they
support the whole weight of the body, and then only for a limited time
do they act as extensors. This will be better understood when we
analyze the movements in the gallop.

There is only one other muscle of the thigh which we will notice,
the *gracilis* (Plate VI., *m*). (It is dissected away in Plate VII.) It is
superficial on the face of the thigh, and is nearly as broad as it is long.
It has its origin on the symphasis of the pubis where it meets its
fellow of the opposite side. It is about an inch in thickness in the
centre, thins off each way, and is attached to the fascia of the leg
for a distance corresponding to its origin at the pubis. It corresponds to the gracilis in man, and is called by Chauveau the *short
adductor*. Its weight is about two and a half pounds. The course
of its fibres is downward and about five degrees outward. In its
contraction the force is as an adductor about ten per cent, but as a
supporter to the weight of the body when it rests on one foot its value
is not to be overlooked.

The want of knowledge of the action of the limbs in locomotion
has led the student of anatomy into a too circumscribed view of the
action of the muscles. It has led him to give first consideration to
forces of secondary importance. It will be seen by a general view of
all the muscles of the haunch that those acting upon the thigh bone,
or femur, from above are inserted on the outer face of the bone, while
those from the lower surfaces of the pelvic bones are inserted into the
inner face of the femur. The primary object in both is locomotion, but,
from the indirect manner of the application of the forces, they are all
necessarily compound ; for example, the great gluteus acts as a propeller and *ab*ductor, while the great adductor acts as a propeller and
*ad*ductor, the *ad*duction of one being compensation for the *ab*duction
of the other. In a humanly constructed machine, as a locomotive,

where the angles are right angles, and the application of power is direct, there is less need of composition of forces; but the design of nature was higher: beauty was superadded to power, and for this end great sacrifices of power were made. Though difficult of demonstration, it may be taken for granted that at full speed the adduction and abduction of all the muscles in action counterbalance each other; if they did not, either the feet would interfere or they could not be brought to support the centre of gravity, and in either case the animal might fall.

How is it possible for the student to learn the action of the machine when the muscular forces are represented as chiefly composed of adductors and abductors, as if the animal was designed to move sidewise like a crab? These names may be perfectly proper to express the action of their analogues in man; for man, of all his relations, has the most inefficient locomotive apparatus, but the greatest diversity of action in his extremities.

Leaving the muscles of the haunch, we descend to those of the leg.

The *triceps femoris* (Plates III., IV., V., IX., *t, t*) is the great muscle that occupies the front of the thigh. As its name implies, it has three heads. The middle one has its upper insertion in the smooth facet of the pubis, directly above the acetabulum, or cup, in which the head of the femur rests (Plate II., 6), and is called the *rectus*. The other two heads are attached to the broad face of the femur, as close as possible to the head of the bone without interfering with its free action. Its length is eleven inches only, but its circumference is twenty, and its weight nine pounds. It cannot be separated into distinct muscles, and it acts as a unit in extending the leg forward through the patella, or knee cap, its point of insertion; but the rectus, or middle head, being attached to the pelvis, has the power of moving the femur on which it lies, as well as of extending the leg in common with its fellows, so that the action is to extend both bones on a line forward; but the patella is not, like that in man, a part of the knee, or articulation, between the femur and the tibia; it has a place of its own. The front portion of the lower extremity of the femur is elevated or built up, and furnished with a trochlea, or grooved surface, with cartilage and synovial membrane, expressly for

the patella to play on as over a pulley; the tendon of the triceps, after
being inserted into the patella, is extended beyond it to be inserted into
a rough tubercle in the head of the tibia. Anatomists call the portion
below the patella, *ligament*. Physiologists may say that the patella is
developed in the tendon. We will not discuss the question. It is to
us as if the bone was developed in the tendon as it is developed in tendinous fibres elsewhere, and the ligament below the patella does the
same office as the tendon above. The force of this powerful muscle
as determined by its circumference can only be compared to the great
gluteus, and is called into action after the extreme of flexion has been
passed, and the femur has been brought forward by its flexors already
referred to, and in which the rectus may have borne a part. After the
foot has taken the ground it steadies the stifle, or knee, and regulates
the flexion of that joint as the angles close to shorten the limb. After
the perpendicular is passed, it again resumes the offensive and extends
the leg in giving the propulsive impulse, which it maintains to the
close of the stride. It rests, therefore, but for one fourth of a stride,
and if the rectus acts as a flexor of the thigh at the same time with
the flexors of the thigh upon the pelvis it has but little rest.

The *gastrocnemii* (Plate VIII., *m, m*), or superficial muscles of the
calf, hold a corresponding position on the leg to that of the triceps on
the thigh, as well as to the levers on which they act; but while the
action of the triceps is very simple and easily comprehended, that of
the muscles of the calf is very complicated, and can only be understood
by a study of the whole limb as a machine of which the voluntary
muscles form a part. Whether it will be possible for me to interpret
the action of the muscles and the use of the tendons with their checks
and reinforcements without the actual limb before us is a question to
be determined. An attempt was made to represent the parts by the
aid of the camera, but the results were not satisfactory. Plate X. is
from a careful drawing by Hahn. The gastrocnemius of the right side,
g, is dissected away from its origin in the femur and raised by hooks
to show the perforatus tendon, *p*. This tendon is inserted into the
femur about two inches from the joint, along with the gastrocnemii
muscles. It has a muscular body of its own, not distinguishable in

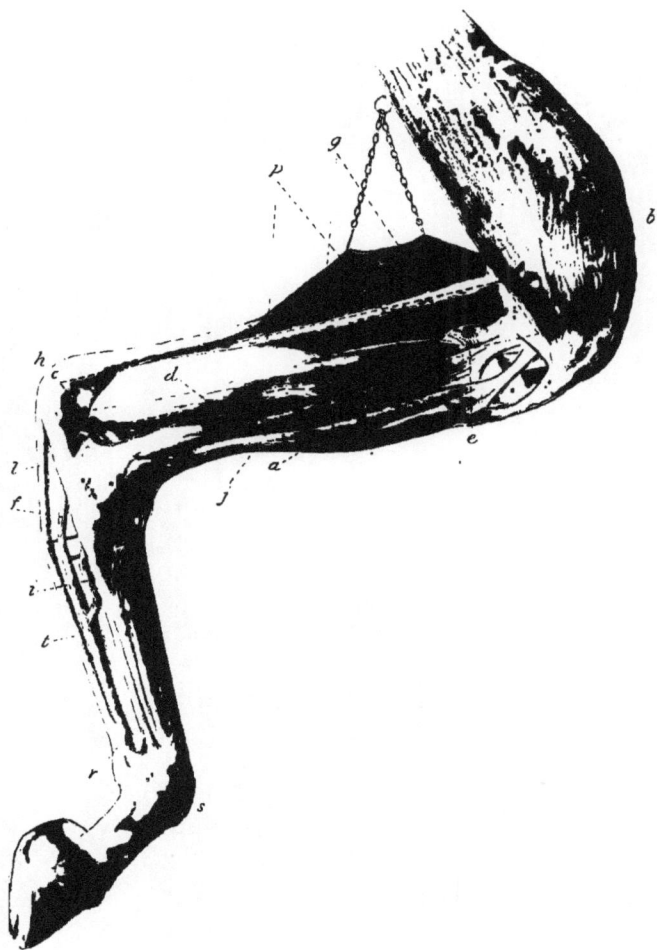

the drawing, being there confounded with the body of the muscle lying upon it. On their way to their insertion into the point of the hock the tendons of these two muscles are twisted upon each other half round, so that the perforatus tendon, which was beneath, reaches its insertion at the outside of that of the gastrocnemius. The tendon of the latter is fixed immovably to the bone, and acts to extend the metatarsus below it, but the tendon of the perforatus passes over the point of the hock, where it is provided with a pulley similar to that at the knee, over which it glides to a very limited extent, being strongly secured by ligaments to the point of the hock, h; it then passes down behind the metatarsus, or cannon bone. to the pastern, or fetlock joint, where it throws out a ring to encircle the tendon of the *perforans*, as seen at r. (These two tendons, forming the "back sinews," would be liable, from the extreme flexions and extensions which take place at that joint, to be dislocated, but for the extraordinary provisions made to prevent it.) It then passes to its insertion into the bones of the foot. When the knee is flexed, as in the plate, this tendon (perforatus), being inserted into the femur above the knee joint, is relaxed, and the extensors of the foot, which are located in front (a), are permitted to straighten or extend the foot, as may be seen in all the plates where the hind foot is in the act of taking the ground ; but when the leg is extended upon the thigh the tendon is drawn upward, and flexion at the joints of the foot is effected, the extensors at a offering no opposition, so that extension of the superior joints, as in the act of propulsion, causes flexion of the inferior. This movement is independent of muscular action, and may be shown in the dead subject, except so far as the act of the extensor of the foot (extensor pedis, a), is concerned ; but the spindle-form body of the perforatus muscle connected with the tendon contracts by volition, and flexes the foot with its added force.

If we consider the limb in the position as given in the plate, and then forcibly extend the foot until the pastern joint, S, is in the position it takes when the horse is standing, the tendon, t, will become tense, and also the ligaments that limit its motion at the point of the hock, h; beyond that it cannot be moved by any force that we can apply short of breaking. It is tied by the ligaments at the apex of

the hock; and if the knee and hock joints are both extended it will
not change the relations, for the tendon, c, m, and the shaft of the
tibia, n, k, being parallel, and the distance from the hock joint, n, to
the apex of the hock, c, and that from the centre of motion, k, at
the knee to the insertion of the tendon at the femur, m, being equal
and parallel, they form a parallelogram, and changes in the angles,
as in flexion and extension, will not affect the length of its sides.
When the knee or the hock joint is flexed or extended, the other
must follow. When the horse is standing, and the knee joint is extended, as well as the hock, the horse rests mechanically upon the
tendons, but the knee is extended by the triceps, b, whose tension
requires an effect of the will and tires in time, so that we see him
when at his ease rest on his hind legs alternately, which he never
does with his fore foot, except when one of them is lame.

The *perforans* muscle, which is so intimately related to the last,
has its origin below the knee joint and on the upper and posterior
face of the tibia and fibula, below the popliteus (Plate X., e), and its
action is not influenced by the flexions of that joint. Its tendon takes
a more direct course to its insertion; it passes through a groove at the
base of the calcaneum, near n, on its inner side and as near the joint as
possible. Strong ligaments cover the groove where the course of the
tendon is changed, to prevent its displacement. It then passes down
behind the metatarsal bone and inside the tendon of the perforatus.
On its course it receives the tendon of another small flexor, and from
the posterior surface of the metatarsus an auxiliary tendon or ligament
of nearly its own size. In the plate this branch is shown relaxed.
The tendon, thus reinforced, is of twice the size it was before the
union, and passes above the pastern through the ring, r, of the perforatus, and is inserted into the bones of the foot.

This muscle, being entirely independent of the femur and the
muscles attached to it, may flex the foot independently, and does so in
propulsion in the last part of the stride, and also in the same contraction aids by its pressure at the back of the hock in extending that
joint, thus extending one joint while it flexes another. When the foot
rests in the standing position, the auxiliary tendon, i, above mentioned

converts the part below it into a continuous tendon, which performs the office of a ligament, in common with that of the perforatus, to aid the suspensory ligament in supporting the weight of the body in the extreme extension which the pastern undergoes when the centre of gravity is over it, as in rapid locomotion.

Muscular fibres are found by anatomists scattered through the tendons below the hock; but for all mechanical purposes the sources of power are above and away from the extremities, where the velocities are, at times, more than twice that of the body and the momentum must be arrested at every stride. The hock in quadrupeds represents the heel in man, and the elongations of bones and corresponding tendons are necessary modifications of the plan for the development of speed.

There is a group of small muscles which form what is called, by some horsemen, the second thigh; they are on the outer face of the thigh and below the stifle, or knee, and in front of the calf. The perforans (*d*, Plate X.) is in this group, occupying the intermediate place.

The *flexor of the metatarsus* has its upper attachment on the tibia, in front of the perforans, and its lower in the metatarsus, below the joint, after passing under the annular ligament. It is minutely described by Chauveau. It flexes the hock joint and is a feeble antagonist to the gastrocnemii, but only acts when the foot is off the ground.

The *lateral* and *anterior extensors* occupy, as their names indicate, spaces on the tibia in front of the latter, and their tendons, after passing under the annular ligament, in front of the hock, descend to be inserted into the anterior face of the foot; they act, therefore, to flex the hock and extend the foot, raising the toe as the limb is thrust forward to take the ground.

The *suspensory ligament* is one of the most wonderful contrivances in the whole locomotive machinery of the horse. Though a ligament only, with its action beyond the control of the will, it is no less an active organ, whose function is indispensable to locomotion, and the interest in it has been much increased by the developments of the camera.

It is not necessary, in order to consider the relations and functions
of this organ, that we should enter into a detailed account of all the
ligaments of the foot; they are very numerous. Anatomists limit the
name to the strong band that has its upper attachment to the metatarsus below the hock, and its lower one into the sesamoid bones, and
they have given the name of sesamoid ligament to that continuation
from those bones to the foot. We will not discuss with anatomists the
question of their identity, but, mechanically considered, they are one,
and, like the patella, the sesamoid bones may be said to be developed
in the ligament. If the name were limited to the first, it would be a
misnomer; for, to suspend the weight that is thrown upon it, it is
necessary that a counter force should act upon the opposite border of
the sesamoid bones equal in strength to that above it. If either part
were divided, the other would have no function, but united they constitute an instrument that often bears the weight of the whole body.
It is a broad, thick band, resembling tendon, and may be felt above the
fetlock between the splint bones and the tendons of the perforatus
and perforans or " back tendons." This ligament fixes the sesamoid
bones in the position above and behind the articulation of the first and
second metatarsals, so that when the second metatarsal or pastern
bone is thrown out from under the first metatarsus they are drawn
into its place, and, their articular surfaces forming an arc of the same
circle, the loss of the pastern is not felt; but the sesamoids now bear
the whole weight of the body, and they have no support but the suspensory ligaments in which they are imbedded, and the tendons of the
perforans and perforatus, which cross the bridge between the sesamoids. The perfect equilibrium between the strength of the ligament
and the force it is required to resist is of the utmost importance.
When the horse is standing upon all four feet, the weight is equally
distributed, and the angles formed by the pasterns with the bones
above are small, for the weight upon each one is not great enough to
spring it far; but in running, the whole weight in every stride is
borne by each foot in turn for a short time, and the elasticity and
strength of its suspensory ligament must be, with that of its reinforcing tendons, just equal to its requirements to support the body, for

they are all placed beyond the control of the will. If it yields too much, the fetlock is liable to strike the ground; if it is too rigid and it does not yield enough, there will be stiffness and a hobbling gait. We shall have occasion to refer to this again when we analyze the paces.

There is no one fact, brought out by the experiments of Mr. Stanford with instantaneous photography, of more interest than the action of the suspensory ligament.

When the horse is standing, it will be seen that the pastern forms an acute angle with the metatarsus. Its position indicates the length of the ligaments, and it is their resistance that prevents the further extension of the joint; but in running and fast trotting, this ligament is put upon the stretch, when the limb is shortened by the weight of the body, to such an extent that the pastern is made to take a position at right angles to the metatarsus and horizontal with the ground. (See the plates of horses speeding, *passim*.) Elongation of the limb begins immediately after the perpendicular is passed, and as the fetlock was the last joint to reflex in shortening, so it is the first to recover its normal extension. This spring continues its action during the rest of the stride, straightening the fetlock joint as the leg becomes elongated after the passage over it of the centre of gravity, still sustaining the body with undiminished force until it leaves the ground, when, being relieved from the superimposed weight, the flexor muscles regain control; and it is the reaction of these ligaments, with that of the flexor tendons acting as ligaments, that produces the quick movement, quicker than is possible in muscular contraction, which causes the feet to throw dirt; it is effected *after* the weight is off the foot and the propulsive effort is complete. There is no muscular action on the foot until after the pressure is removed and the flexors regain control.

It is an exceedingly difficult problem to determine the absolute, or even the relative, work performed by the different muscular powers employed in locomotion. There are many different elements entering into the calculation, that are impossible to be weighed. Muscles differ in quality as well as quantity; some contain a larger proportion of cellular or fibrous tissue than others, and will have less power, other things being equal. For example, the gluteus and vastus are coarse muscles capable of resisting external force, and therefore popularly believed to be strong; but it is in a meaning corresponding to toughness, and that quality depends upon the amount of interstitial cellular tissue they contain, which tissue has no contractile property, and cannot originate motion; while the psoas and iliacus, having but little such cellular or fibrous tissue, have little power to resist external force, but have a larger contractile power as measured by the areas of their sections.

Muscles do not often have their force concentrated at both extremities, but it is distributed over the face of their levers at different distances and at different angles, as in penniform muscles, and nearly all others in a greater or less degree, and at different angles at each change in the position of the levers. Though we recognize the same general mechanical principles, we cannot apply the same mathematical rules usual in mechanics; add to these elements of uncertainty the composition of forces often in the same muscle, and we see how formidable are the difficulties in the way of reducing animal mechanics to an exact science.

But while we cannot accurately determine the forces in detail, we can in the aggregate. We see all these different and often antagonistic forces united in their action around a common centre of motion, as the hip joint, to effect one result. There are certain general principles, however, that we can deduce from the facts before us. In order that the foot shall reach the ground as far in advance as possible, to support the centre of gravity as early as may be, and as long as possible, and that it may use its propulsive force later, it is necessary that it should be possessed of sufficient length; but it is bearing a burden whose weight we will suppose to be a thousand pounds, and going at the rate of twenty miles an hour, and the momentum is the product of that weight multiplied by the velocity. This is a responsibility that could not be borne on stilts. The difficulty is overcome by so constructing the whole limb that it shall be extensible, thus having all the advantage of length without its disadvantage; and the centre of motion is actually lowered several inches that its practical length may be increased. For this purpose the system of levers is used, which, by their flexion and extension, practically shorten and lengthen the limb. The acuteness of the angles at which these bones intersect each other is, therefore, an important element in the mechanical action ; the angles to be acute require long levers, and long levers necessitate long and powerful muscles to "man" them. These qualities must be bred.

Flexibility of articular ligaments may be acquired by early training and regular exercise, but the proportions of the body are inherited. Length of muscular fibres and acute angles of the levers on which they act, give sweep of limb, and strength depends upon the number of them, and the effective power of both depends upon the will or courage ; but all these qualities would be vain if the motion of the extremities were not so co-ordinated that their functions should be performed without interference one with another.

When the speed of the horse is twenty-five miles an hour the rate of the hind foot in passing that on the ground is twice that, or fifty miles an hour. It is even greater than that, for the velocity of the foot in its stride is an accelerated one during most of the distance, and may be supposed to be most rapid midway. Now the movements of the posterior extremity on its centre are controlled by voluntary muscles, liable from various causes to be irregular, as they must necessarily be from the ever-changing centre of gravity which it is designed to support. There would have been danger of one foot striking the other leg in passing, — an accident technically called interference, — but another danger still greater existed at the stifle from the blows that joint would be liable to give the abdomen in its extreme and violent flexions. It is the duty of the iliacus muscle to guard the abdomen from this violence, and when it performs its office well, it gives the "stifle action" so much admired ; but while the upper end of the leg (tibia) is thrown out in this action, the lower end is correspondingly thrown in, and the foot would be still more so but for the unique construction of the hock joint. The interlocking grooves of this joint are not direct, as in other hinge joints of the body, and as the corresponding joint in man is, but oblique, so that when flexion takes place at that joint, the lower ray is carried obliquely outward, and when the other leg is passed, and the extension takes place again, its action is reversed, and the foot is returned to the position required to support the centre of gravity. By this simple contrivance the danger of this accident is placed beyond the will of the animal, and in well-formed horses beyond the possibility of accident. Some horses circumduct the hind feet more than others, and in others the stifle action is most marked; but it is not common to see both excessive in the same horse.

There is often considerable difference in different horses in the length of the hock. The long hock gives the greatest power, for the reason that the leverage is greater; but what is gained in power is lost in speed.

Sometimes there is a looseness in the articulations of the tarsal bones immediately below the hock joint, which, by their freedom of motion upon each other, enables the joint to become more extended, and the last effort of the gastrocnemii muscles is given with great advantage of mechanical power from the practical shortening of the arm of the lever on which they act, and from the ability the limb acquires of retaining its position upon the ground for a longer time. It is a point in some fast animals, but would be considered a defect in a draught horse.

Having given a detailed description of the parts concerned in the motion of the posterior limb, and their action, I will now endeavor to show how the machine acts as a whole. If the reader has familiarized himself with the parts by reference to the plates, while he has followed the description, he will experience no difficulty; but if he has not, it would be as well for him to pass over the rest of this chapter. The analysis has no reference to any particular gait or co-ordination of the limbs with each other, but it is confined to the action of one posterior limb alone, and it will be found to be the same in all the paces, differing only in the degree of action according to speed.

We will take for our guide the posterior extremity as it has just left the ground, after the act of propulsion is complete, and in the medium pace, the trot.

In order to aid the mind in understanding the actions of the muscles upon their levers, the skeleton is mounted with movable joints, by which means we are enabled to adapt it to every position required. By this means it is a comparatively easy matter for one to understand the action throughout. (See Plates II., XIV., XV.)

Retraction begins by the relaxation of the gluteus maximus, the vastus, semi-membranosus, and the great adductor. The triceps also relaxes, and the tibia is free to respond to the contraction of the semi-tendinosus lifting its lower extremity. The tensor vaginæ, acting from the hip upon the knee, the psoas magnus, iliacus, and sartorius from the inner and upper wall of the pelvis, with the anterior branch of the superficial gluteus from the hip, all act in concert to advance the thigh, the knee becoming more flexed as it is advanced; and with the knee, or stifle, goes the hock joint, by the relaxation of the gastrocnemii and the mechanical arrangement before described.

The flexors of the foot act at the instant their tendons are released from the forced service as ligaments, and continue their action until the perpendicular from the centre of motion to the ground is reached, which marks the point of greatest flexion of all the joints. The flexors of the thigh, already mentioned, maintain their tension to keep the lower extremity of the femur in its advanced position. The semi-tendinosus relaxes, while the triceps extends the tibia upon the femur already well thrust forward, and the muscles of the calf, acting on the point of the hock, extend the metatarsus synchronously with the feeble action of the extensors of the foot. The perforans and perforatus do not take part in this movement, as their action would counteract that of the extensor. In this order the foot takes the ground, the heel being the first to make the contact, and by its elastic frog it is peculiarly fitted to receive the shock. It will be observed, by reference to the plates, that the bones of the entire limb are at angles best adapted to meet the contact with the ground. The toe is raised to avoid tripping, and allow the elastic frogs of the foot to make the first contact.

The instant of contact, when the foot is as far forward as possible to sustain the centre of gravity, marks a sudden change. The flexors of the thigh, the sartorius, tensor vaginæ femoris, iliacus, and the

8

anterior branch of the superficial gluteus, give way, while the weight of the body relieves the extensors of the foot. The function of the limb at this time is to support the weight of the body and prevent it from pitching headlong; and to this end, with the exception of the few small muscles just mentioned, the entire mass of the muscles of the limb is called into action; and now that the foot is a fixed point, the semi-tendinosus acts in unison with the others to take the weight of the anterior half of the body. This is the use of all the vast mass of muscular power developed in the haunches and long muscle of the back (ilio spinalis). In this manner there is no act of extension, further than the extension of the body upon the thigh; it is not until the centre of motion, or head of the thigh, has passed over the foot that extension is possible; and then the nearer to a horizontal the direction of the force applied, the more effective it will be. When the limb is perpendicular, the whole force is employed in supporting weight; but when it is exerted upon the ground at an angle of forty-five degrees, one half of the force is spent in supporting weight, and the other in propulsion; if it could be exerted horizontally, it is plain it would be exclusively spent in propulsion. From the time when the foot is planted in advance, until the leg has passed the perpendicular, the force is also compound, a part being employed in supporting weight, and the other in resistance which must be drawn from the momentum; this last is reduced to the minimum by the gradual giving way of the triceps and gastrocnemii, and contraction of the great propellers of the haunch, especially the vastus, which forces the trunk over the supporting limb. The act of propulsion by the vastus begins from the moment that the hind foot takes the ground and its contraction begins. The effect of the contraction of this muscle is to shorten the distance between its two extremities; one of these extremities is attached to the lower end of the femur and the other to the spines of the sacrum behind the croup, but the course of the muscle is not direct (see Plate V.), being deflected at the head of the femur, and most so when the foot first reaches the ground. At that time it presses with most force against the articulation pressing it forward, so that it extends the trunk upon the limb and forces it forward in the same act.

After passing the perpendicular, and the angles of the extremity are increased, the semi-tendinosus ceases to act, and the extension is continued by the vastus, gluteus, triceps, and muscles of the calf, to the end of the stride. In the flexion of the limb that takes place as it shortens in order to give uniform support, and not be itself crushed, the flexion is effected by the weight borne, in which the flexors proper bear no part; their action could have no other effect than to bring the body to the ground, but it is effected by the gradual giving way of the triceps and the suspensory ligament.

It will be seen that but a small part of the immense power of the extensors, or propellers of the posterior extremity, is spent in the act of propulsion, even when the animal is in full motion, but in supporting weight; and as the extension of the leg increases and the burden is assumed by another limb, it is the better enabled to exert its propelling power. As the limbs are successively relieved of that duty by their alternates, they are in better position to exercise their functions as propellers.

This analysis of the mechanism of the posterior extremity will become of importance when we come to apply it to the run or greatest speed of the horse. The reader who has not had the patience to follow us through the study to the end of this chapter will not be able to master the next, and we would advise him to pass it over, and take up the fifth chapter, where we will endeavor to apply the demonstrations contained in these two; but such must take the facts on which the theory of motion is based for granted.

CHAPTER IV.

THE anterior extremity furnishes a subject for the study of animal mechanics of more interest even than that which has demanded our attention in the preceding chapters.

There appear at first sight greater difficulties in the way of human ingenuity in the application of mechanical power for propulsion to the anterior part of the trunk. The mind is led by the similitudes of comparative anatomy, and the popular hypothesis of evolution from one common parentage, to look upon the anterior extremities as limbs in progress of development into arms or tool-makers. The mind jumps, like the kangaroo, from the marsupials to the monkeys, to the orang-outang, and then to man by such easy leaps that it is difficult to persuade one that he has advanced to his opinions without substantial grounds. To these causes must be ascribed the

universal opinion of writers on the horse that the fore legs are
merely supporters; and the latest and standard authority on the
horse, in England, compares them to the spokes of a wheel, and asserts
that their only functions are to support the centre of gravity and keep
out of the way of the propellers, the hind legs.   It will be apparent
to the reader before the conclusion of this chapter, if it is not so
already, that each limb is required to support the body and act as
propeller in turn, and that the anterior one does more than its share
of both offices.*

It will be shown, when we come to analyze the fastest pace of
the horse, that the strongest propulsive force of either of the legs
is given with the anterior one in each stride; indeed, it is so strong
as to raise the centre of gravity several inches above the horizontal
line of its motion.   As the case now stands between the anterior
and posterior extremities, they may be compared to a peasant and
his wife in certain foreign lands, in which the latter is required to
share equally with her husband in all his labors and also to bear
burdens which he cannot share with her.

The beautiful contrivances by means of which the anterior limb
is enabled to support weight as a crutch, to be acted upon as a
passive instrument in propulsion, and at the same time to consti-
. tute an autonomy of its own, independent of both the others, for the
accomplishment of the same general result, cannot fail to excite the
most profound admiration, and wonder that its mechanism has not
been better understood.

On reference to Plate IV., *s*, one will see the posterior half of
the *great serratus* brought into view by the removal of the superficial
muscles that hide it in Plate III.   It is so called because its lower
border is serrated or notched, the lower attachments being to the
first eight ribs; the anterior half of the muscle is concealed by the
shoulder.   This muscle is fan-shaped, its fibres converging upward

---

* Mr. Walsh (Stonehenge) gives the authority of M. Baucher for the statement that
the weight borne by the anterior and posterior extremities, as determined by placing them
upon different weighing-machines, was as 210 for the former to 174 for the latter, the total
weight of the horse being 384 kilogrammes.

to a common centre on the inner face of the upper border of the
shoulder-blade, or scapula, as seen in Plate XI., *s, s.*

When this muscle is recently exposed it presents delicate nacreous tints rivalling pearl. The artist has suggested them only in his
drawing. This pearly coat of the muscle is tendinous in its structure, and extends over the whole exterior surface of the great serratus. These tendinous fibres extend throughout the muscle, but are
in greater proportion near the centre or long axis (Plate XI., *a*).
These tendinous fibres, concentrated at *a*, may be considered the
centre of motion for the whole limb when supporting the weight of
the body, whether acting alone, or in conjunction with one or more
of the other limbs, and whatever may be the direction of its axis
with reference to the trunk; but this centre of motion must not be
confounded with the centres of motion existing in the joints; it
holds a corresponding position with the "whirlbone," or hip joint
of the posterior extremity. This intermixture of muscular and
tendinous fibres existing in this muscle is found in others, as the
deep gluteus described in the last chapter, enabling it to perform
the functions of both muscle and ligament. The tendinous fibres,
which are in the greatest proportion in the long axis, when put
to their tension absolutely limit elongation to that degree, and are
useful when the animal is standing; as these tissues are incapable
of fatigue, so he has no occasion to rest them. With the aid of
another muscle, which we shall describe further on, having the same
characteristic construction as the serratus, the horse is enabled to
stand in his stall all day without resting either of his fore legs;
while in the hind leg the labor falls upon the triceps (Plate IV., *t*),
of pure muscular fibre, and he will be observed to rest his hind legs
alternately. (See page 50.)

The muscular fibres of the serratus are most abundant at the
anterior and posterior borders. The former aid in preventing shock
when the foot first takes the ground, and the latter in giving the
final propulsive effort when it leaves it; and by their joint action
they relax the tendinous fibres, or bands, which, being passive, have
no such power in themselves.

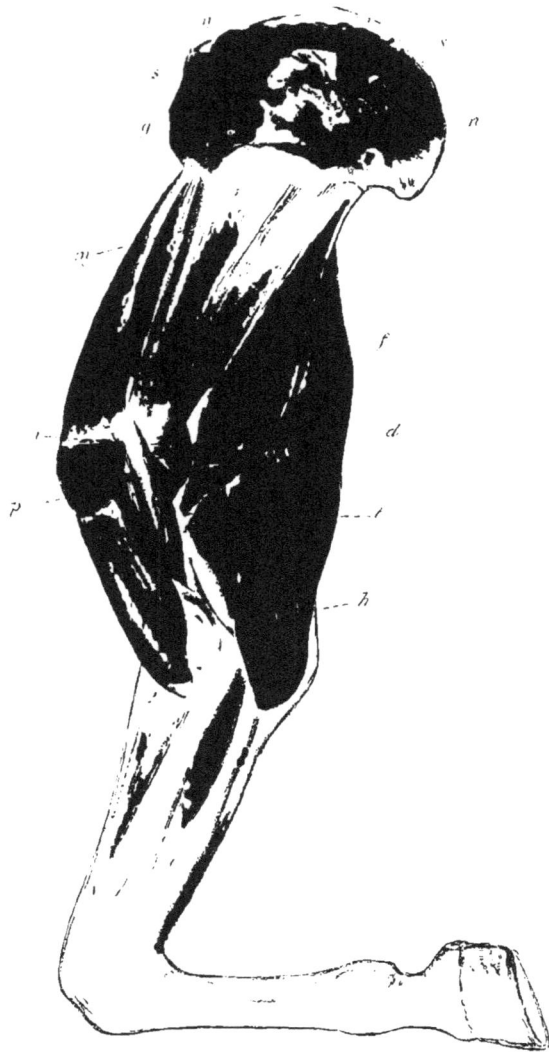

The centre of motion in the anterior extremity may, in its mechanical function, be considered as a joint, and the only kind of joint possible in that position; were it constructed like the corresponding joint in the posterior extremity, it would be inevitably broken by the contact with the ground, thrown out as it is in advance of the centre of gravity. For the same reason it is not provided with a collar-bone, or clavicle, as in man and the anthropoid animals, in whom that bone fixes the shoulder and makes it the centre of motion for the limb.

On reference to Plate IV., s, the great serratus will be seen as a fan-shaped muscle which has its lower attachments spread out over the first eight ribs. From the attachment to the different ribs its lower border is like a saw, from which its name, "serratus." The artist has vainly attempted to represent the nacreous color, in which it vies with the mother-of-pearl. This is the tendinous covering to the muscle, and it is much intermingled with tendinous fibres, which limit elongation and take the strain from the muscular fibres when their contraction is not called for. The upper attachment of this muscle is on the inner face of the scapula, or shoulder-blade (Plate XI., a), below the cartilaginous border, with the dark line marking the boundary between it and other muscles. In the centre are seen the gray fibres of tendon, which are continuous below, and enable the animal to rest the muscular fibres and limit their elongation. The space covered on the inner face of the scapula is nine inches in its greatest measurement by two in its least.

The space below the section of the serratus, as seen in s, s, Plate XI., and between that muscle and those of the inner face of the shoulder-blade, is lined with loose cellular tissue, which, while it connects the opposing surfaces, allows of unrestricted motion upon the centre, a, and prevents friction. The body in a standing position rests the weight of the anterior half upon these serratus muscles as upon a sling to which the anterior extremities correspond to crutches.

But when the foot of one of these limbs is off the ground the serratus is relaxed, and the limb would drop but for another set of muscles, which, though feeble, are sufficient for the purpose which

they serve. This is the special function of the *trapezius* (*g*, *g*,
Plate III.). It is so perfectly represented in the plate that it requires
but little description. It is divided into two parts by the spine of
the scapula (see Plate II.), into which both divisions are inserted
along with a band of the ligament of the neck, which seems to be
sent off for the purpose of aiding with its passive force the trapezius in holding the limb to its place. The upper insertions or origins
of both divisions are in the same ligament of the neck, or *yellow cord*,
as it is well called by hippo-anatomists. This cord is distinguished
not only by its color but by its elasticity from all other ligaments.
It seems to be, indeed, a special contrivance to afford means for the
attachment of important muscles when the spines of the vertebra
are too remote to afford it. It extends from the head to the strong
spines of the dorsal vertebra, where it becomes merged into ordinary
ligament. It may be that the branch of this cord that is inserted
into the spine of the scapula is itself sufficient to support the weight
of the anterior limb, and that the muscle under consideration is used,
the two parts acting alternately, to aid in locomotion, exerting their
force at the upper or cartilaginous extremity of the scapula and above
the centre of motion or attachment of the serratus; but however that
may be, its aid in locomotion cannot be great, as its entire weight
does not exceed two ounces. Its thickness does not vary much from
half an inch. It is separated from the skin only by the general
aponeurosis, or fibrous covering described in a former chapter, and
which has been dissected away from the whole body in the subject
of the drawings.

When the trapezius is removed, the *rhomboideus* is brought into
view. This muscle is so named from the corresponding muscle in
man, in whom it is in the form of a rhomboid; and if the name were
limited to the muscle so far as it corresponds to that in man there
could be no objection to it, but since Cuvier's time it has been made
to embrace another muscle, the *levator anguli scapulæ* (Plate IV., *r'*).
To this union in the horse there can be no objection, anatomically
or mechanically; but when so united they are no more like a rhomboid than a tent-pin, and the name of levator anguli scapulæ should

THE HORSE IN MOTION. 65

have been applied to the united muscles, if either; but the worst
part of the history is that the name of levator anguli scapulæ was
applied to another muscle, the *trachelo subscapularis* (Plate IV., *g, g*).
No name could be more inappropriate than this; in no way, directly
or indirectly, can it be said to lift the angle of the scapula, as may be
seen by reference to the plate. The function of this last-named muscle has, so far as I know, never been understood until now, and will
be explained further on. But this furnishes another example of the
confusion arising from hippo-anatomists being misled by human anatomy. The levator anguli scapulæ is quite distinct from the rhomboideus in man, having its origin in the transverse processes of the
vertebra of the neck, while in the horse its origin is in the spinous
processes of the vertebra, as far back as the withers and along the
yellow cord. (See Plate IV.) In man, its name, *lifter of the angle
of the scapula*, is good, for that expresses its function; in man,
however, it is no locomotive organ, but even more necessary to
the complicated movements his superior extremities are required to
perform.

From the necessity which exists, for the reasons given, of restoring
the old name to the trachelo subscapularis, the restoration of the
name levator anguli scapulæ to its old association becomes necessary,
if it is not to be abandoned altogether. In order to make intelligible a description of the mechanical action, there is need of definite
terms, and we will apply the name levator anguli scapulæ to include
the rhomboideus as well.

The contraction of its fibres does not take place until the leg is
extended and the foot rests upon the ground; it then acts to draw
forward the upper or short end of the whole extremity as a lever
with its fulcrum on the ground and its weight at the centre of motion.
The course of its fibres is accurately drawn in the plate, the limb
being in its normal position. Its posterior fibres are few, but as it
extends forward they become numerous and more powerful. Their
insertion is into the inner border of the cartilage (Plates IV. and XI.,
n, n, n'); at the anterior border, *n'*, is the insertion of the muscle known
before the time of Cuvier as the levator anguli scapulæ; they form a

9

considerable mass, and join on to the serratus, s, so nearly in the
line of the centre of motion that it may be that they act in conjunction with the trachelo subscapularis, whose insertion is at g,
Plate XI.

The last-named muscle is well exposed in Plate IV., g, g. As
there seen, it is triangular. It arises from the transverse processes of
the last six cervical vertebra, and its fibres converge to their insertion on the inner face of the scapula, in front of the insertion of the
serratus, or centre of motion. Its muscular fibres are in little fasci-
culæ, or bundles, separated by interstitial fibrous or cellular tissue,
to admit of great freedom of motion upon each other in the extreme
vertical flexions of the neck while grazing. It is a powerful muscle,
its weight being three and a half pounds; but its action has not been
comprehended, its fibres being nearly horizontal on an average, or a
little upward, and their insertion on a line with the centre of motion;
it can have no active agency in locomotion, though with the joint
action of the levator anguli scapulæ it may move the upper end of the
scapula forward, as far as permitted by the tendinous fibres of the
serratus and the branch from the yellow cord; but that cannot be
much. Until the theory of quadrupedal motion was understood its
function may well have been overlooked. It is now clear. Its attachment being on a line with the centre of motion and directly upon a
fixed point, it cannot be supposed to aid in the motion of the scapula
about that point; but when the animal is running, and the fore leg
is thrown forward and takes the ground, it is required alone to
receive the weight of the whole body or be itself crushed by its
momentum. This will be resumed after the action of the *triceps
brachii* at the same instant is shown. It is sufficient for the present
that the action of this muscle abstractly be understood and remembered. Its general appearance is so like the muscle above it, the
scalenus (*m*, *m*, *m*), having similar origins along the cervical vertebra nearer the head, and its insertions into the spines of the dorsal
vertebra (hidden in the plate by the overlaying levator anguli scapulæ), that it is apt to be confounded with it; but the scalenus is
not a locomotive muscle, its function being to raise the head when

the pair act unitedly, and to bend the neck laterally when each muscle acts separately.

We have shown how the anterior extremity is used as a supporter to the trunk, or crutch, and how it is itself supported in its position when not so acting. The mechanical principles involved are very simple. The method in which mechanical power is applied to the same limb as a lever in locomotion will be found to be no less so, and if the contrivance does not display as great ingenuity as some parts of the locomotive organs, it is because there was no occasion for such display: it has the merit, at least, of being very primitive.

While there is no bony connection between the anterior extremity of the horse and its trunk, therefore no fixed point of resistance and reaction, as in the posterior extremities, the centre of motion is attained equally well, and it is difficult to conceive how it could serve its different relations to the trunk any better. The centre of motion in the anterior extremity is in the scapula, as high as a bony base could be reached. This, if not anatomically so, is mechanically a joint, and corresponds to the hip joint of the posterior extremity, the shoulder to the stifle, and the elbow to the hock. In this view, there is no reversed order in the joints, as has been stated, but the same mechanical relation. The freedom of motion at its centre in the limb is less than in the corresponding joint in the posterior extremity, but there is all that is required; it is placed considerably higher than in the latter, in order that more motion should not be required; and the restriction at that point is compensated for by the superior flexibility of the lower joints. The total result is that the stride of one limb is just equal to that of the other.

The limb, acting as a lever of the third order, having its centre of motion as high as possible, should have the power to move it applied as low down as possible, within the periphery of the body; but the farther from the fulcrum, or centre of motion, the power is applied, the greater will be the space moved over, and, consequently, the longer must be the fibres of the muscle.* This requisite is furnished by the *great dorsal* (Plate III., *d, d*), which has for its base the spines of the last

* See page 31.

fifteen dorsal and the lumbar vertebra. Though spread over so much space, the muscular tissue is not correspondingly extensive. The purpose was to gain advantage of position as far back as possible to give the most direct action in the line of motion to be produced. As the fibres of its thin tendon (Plate III., *f, f*) converge forward and downward, they become more muscular, and most so just behind the scapula, which is covered at its posterior angle by it, and it is covered in turn in the same region by the dorsal division of the trapezius, *g ;* after passing beneath the muscles of the shoulder (as seen in Plate XI., *d,*) its fibres again change to a thin, flat tendon, which unites with the tendon of the muscle, *f,* and is inserted with it into the internal tubercle of the humerus, about one third of the way from the shoulder to the elbow. If this muscle acted when the foot is off the ground, it is plain that it would flex the shoulder; but its function as a propeller is called into play when the foot is the fixed point, and the limb is supporting the weight of the body, and its articulations are all set. Under such conditions it forces the body forward over the foot; but its power as a propeller is second to that of the *great pectoral* (Plate III., *p. p*). The limits of this muscle are a little in doubt. It is represented in the plate with the boundaries as given by Chauveau, but it is confounded so closely with the superficial muscle of the skin (*paniculus carnosus*) on its upper border that it is difficult to separate them. For our purpose, it is sufficiently shown in the plate, extending from the tenth rib over the thorax, covering the *serratus magnus* as high as the lower border of the great dorsal, and as low as the middle of the thorax, where it unites with its fellow of the opposite side. Its fibres converge as they are directed forward, and form a mass of muscle between the arm and thorax so great as to be second in power to no other locomotive muscle in the body. Its insertion is into the inner tubercle of the head of the humerus, as seen at *p,* Plate XI., as near to the shoulder joint as possible. The great dorsal may perform two functions, flexion or propulsion, as mentioned. The muscle now under consideration has but one. Acting directly upon the angle of the shoulder, there is no loss of its immense power by indirect force, and from the moment that the foot touches the ground, its power is felt in forcing the body over it. As there is no loss of force in indirect action, so there is none spent in adduction or abduction, or in supporting weight; that office is performed by the muscles of the limb acting automatically, and the effect of its traction upon the shoulder is to support it and prevent it from giving way while the limb is playing its independent part in sustaining the superimposed weight of the body.

There seems no room for a doubt that the conjoined action of the two sets of muscles last described is the most powerful propelling force in the whole locomotive organism of the horse. To make this apparatus complete, there was necessary some force to return the limb to its position forward when the act of propulsion was completed. This force is found in two sets of muscles, the *mastoido humeralis* and the *superficial pectoral;* the former has its fixed insertion at the mastoid process of the temporal bone, or base of the skull, behind the ear, and to the first four cervical vertebra.* It is shown in Plate III., *m, m, m,* passing downward and backward along the whole length of the neck and over the point of the shoulder, enveloping it, and is inserted at the humerus, about half-way from its two extremities. (See Plate IV., *i,* where the muscle has been cut away from its tendon of insertion. It is also severed at *j,* leaving only its upper portion *in situ*.) It is six inches in width where it envelopes the shoulder joint, and an inch in thickness, and gives off, about thirteen inches above its insertion, a branch to be inserted at the anterior border of the sternum, or breast-bone. This branch, which could not be well shown in the drawing, is known to anatomists as the *cuticularis colli.* There does not appear to be any occasion to consider it a distinct muscle; its fibres are interwoven with those of the muscle under consideration; its function is to aid that muscle, and fix it in its position over the shoulder joint. Though so thin, the weight of the mastoido humeralis is not less than five pounds. To give effect to this muscle, it is necessary that its base, the head, should be fixed. This is effected by the *complexus,* and its allies of

---

* This relation is not well shown in the drawings, owing to displacement, caused by the cord used in suspending the subject; the artist drew the parts as he saw them, and the inaccuracy was overlooked until too late to be corrected.

the neck. From this it follows that the horse, in speeding, should be allowed to follow its instinct in fixing the position of the head. The ally of this muscle is the superficial pectoral, which has its insertion on the anterior extremity and lower margin of the sternum, or breast-bone. The course of its fibres is backward, downward, and outward ; they divide into two branches: one is inserted into the anterior ridge of the humerus, along with the mastoido humeralis ; the other is spread out on the fascia of the inner face of the leg. The action of this muscle is to carry the whole limb forward, in common with the last described, and at the same time to adduct it to counteract the abduction of that muscle.

The action of these two sets of muscles is so unlike any other that it is not readily understood. Let us suppose a man propelling a boat through the water by means of an oar, and the handle end of the oar made fast to the side of the boat opposite to that on which he is seated, but free to move about a pin ; then let the man remove the rowlock from its place and substitute for it his hands ; next make fast the blade of the oar in the water, and the man shall then apply his strength to the oar: the boat will move. Now, if this illustration be modified so that the oar shall be vertical, and the blade of the oar be fixed to the bottom, and the handle to a fixture above the man's head, the similitude will be complete. Of course the nearer the power is applied to the foot, or fulcrum, the faster the upper end will move, but the greater must be the expenditure of power. There is another muscle, acting from without upon the shoulder, whose office has been doubtful, the *small pectoral.* It arises from the keel of the sternum, or breast-bone, and passing between the shoulder and the neck, fills the angular space in front of the scapula. It is thick below, where it is turned over the breast, and becomes smaller as it is reflected on the scapula, tri-angular in form, to fit the space it fills, and is the muscle against which the collar rests ; this is a muscle of considerable power, being two and a half pounds in weight. It is attached to the muscles of the scapula by strong cellular tissue, and to the strong aponeurosis that covers it. Besides being an element of beauty, by giving graceful contour to the parts, it seems to have no other function than to pull forward the whole limb, rendering tense the tissues connecting it with the trunk, and by so doing extending the limb to enable it to take the ground farther in advance, and leads us to infer how great importance was attached by the Master Mechanic to utilizing every available means to enable the fore foot to reach the ground as far in advance as possible, that no time might be lost in giving support to the centre of gravity.

We have thus far considered the anterior extremity as a passive tool taken as a unit; it remains to study it as an active automatic machine. It is difficult to trace any analogy between the mechanism of the anterior and posterior extremities thus far; but in the system of levers, by the closing of which the limb is shortened, and in the opening of which it is lengthened, we recognize the same mechanical combinations that are employed for the same purpose in the posterior extremity.

In Plate IV. the external view of the shoulder and arm is given showing its relation to the trunk and that of the muscles to each other. The pearly-colored upper border of the scapula, _n, n_, is seen with the levator anguli scapulæ still attached. This border, which is cartilaginous, is not seen in the prepared skeleton, but a rough margin to the bone indicates its former connection. No muscle of the anterior extremity, as an automatic machine, is attached to this cartilaginous border. It has not sufficient firmness to resist force from below, but its tenacity is sufficient to withstand great traction, and its flexibility is such as to prevent any danger of fracture by force so applied. The spine of the scapula may be traced from the cartilage downward, near the middle of it, to which the trapezius and branch of the yellow cord were attached. This spine divides unequally the scapula; in front of it is the _superspinatus_ muscle, _s s_, whose terminal tendons pass over the head of the humerus, or shoulder, one to be inserted into the external tubercle on the outside, and one third of the distance from the point of the shoulder to the elbow. This is the most considerable division, and acts to extend the humerus on the scapula and rotate it outward. The other tendon is inserted near the internal tuberosity; it unites with its fellow in extending the humerus. Considering this muscle mechanically, it would be proper to regard its

lower insertions as one, overlaying the joint beneath the mastoido
humeralis, and acting on the head of the humerus as a direct extensor
of the humerus; its weight is two pounds, and its length seventeen
inches.

The shoulder joint is constructed on the same principle as that
of the hip, but the head of the humerus is broader and less convex,
and the cavity of the opposing articular surface of the scapula too
small to lodge it; but it is supplemented by cartilage and ligaments,
and held still more strongly in its position by the powerful tendons
which envelop it.   The head of the humerus is held in its place
by the further assistance of the atmospheric pressure equal to one
hundred pounds.   Though freedom of motion is not so great as in the
corresponding articulation in man, it is much greater than that of
the hip joint.

The two muscles that especially guard the joint and prevent lateral
displacement are: the *infraspinatus* (*i s*), which is attached to the external surface of the scapula, and nearly fills the space below its spine.
It is inserted into the head of the humerus, at *o*, directly opposite the
shoulder joint; the other is the *subscapularis* (Plate XI., *m*), having
its attachment on the inner surface of the scapula, and occupying the
whole face of the bone below the insertion of the serratus, *s*, and it
is inserted into the inner side of the head of the humerus, directly
opposite to the insertion of the infraspinatus.   These two muscles
are of the same power, each weighing two and a half pounds, and of
the same length.   Acting simultaneously, they neither flex nor extend
the humerus, the abduction of the one cancelling the adduction of
the other, but they are powerful braces to the joint.

There is another pair of muscles, whose functions cannot be understood unless considered together.   If the reader will refer to any one
of the silhouettes of the trotting horse, and watch the action of the
fore leg from the time that the foot leaves the ground until it takes a
new position in advance, he will perceive that all the joints are flexed
rapidly before the foot passes the perpendicular.   The flexion at the
shoulder is performed by these two muscles.   One is called, by Chauveau,
the long adductor; *teres minor*, by Percivall; and the *scapulo hume-*

ralis, by Legh. The other is called the adductor of the arm, by Chauveau ; *teres major*, by Percivall ; and the *great scapulo humeralis*, by Legh. One acts on the outer and the other on the inner aspect of the humerus, at equal distances from the shoulder joint, and nearly one third of the distance from the articulation ; one from the outer, and the other from the inner surface of the scapula; and the weight of each is one pound, while their length is the same. They cannot be conceived as acting independently of each other, and it is useless to consider what their function would be when so acting. Conjointly they are neither adductors nor abductors, but flexors of the shoulder. While the bone is thus flexed, the limb is brought forward by the mastoido humeralis, which is inserted into the same ridge as the external of these two muscles.

When the time comes for a thorough revision of the names of the muscles of the horse (and that time must come soon, for it is now confusion worse confounded), it is to be hoped they will be determined by their mechanical action without reference to the action of corresponding muscles in man. The camera has now made the task comparatively easy. When that time comes, these muscles should be known as the *flexors of the shoulder, internal* and *external.*

There are two flexors of the forearm. The *flexor brachii* is a short tendinous muscle, originating from the lower anterior extremity of the scapula, just above the centre of the shoulder joint, by a strong tendon, which is developed into a patella-formed cartilage, moulded to the double groove on the anterior angle of the humerus, over which it glides as a synovial articulation, or a pulley, in the same manner as the patella of the stifle joint, the grooves being deep so as to prevent lateral displacement in extreme flexion. Below the shoulder it forms a cylindrical muscle ten inches long. Its muscular fibres are intermingled with tendinous bands, by which its elongation is limited, and it is enabled to act as a ligament to support the weight of the body without fatigue.* It is inserted into the capsular ligament of the elbow joint, and the rough tuberosity at the head of the radius. It raises the forearm, and is one of the muscles on which, in part, the high

* See description of serratus muscle, page 62.

10

action of the knee depends. The other is the *humeralis externus* of Percivall. It originates behind and below the head of the humerus, and, winding around that bone, fills the furrow of torsion (Plate II., 34). It is inserted into the anterior heads of the radius and ulna, and acts as an assistant to the flexor brachii: the two muscles originate from opposite sides, but act as a unit lifting the forearm. It has greater power than its associate, being larger and more muscular, and from its spiral course its fibres are longer; it is capable, therefore, of giving higher action than its associate.

The *triceps of the arm* is a powerful muscle which plays a very important part in the mechanism of the anterior extremity. As its name implies, it is a three-headed muscle, if we choose to consider it one muscle, and it is an extensor; but the correspondence in name with the triceps extensor of the thigh should not lead us to confound its mechanism with that of the latter. The triceps of the arm (Plate IV., *h*) fills the angular space between the point of the elbow (olecranon process) and the lower border of the scapula. The infra-spinatus, *i s*, covers the origins of the three heads, but their common insertion at the short end of the ulna, as their lever, is clearly shown in a strong tendon. The two upper heads are attached to the lower border of the scapula, and when these divisions contract they tend to close the angle between these bones; but the third, or lower head, is not attached to the scapula, but to the posterior face of the humerus. This branch, sometimes called the short extensor, being independent of the scapula, may act in extending the arm when the angle formed by the latter bone and the humerus is so small that the limit of contraction of the other two branches is reached, as is the case in every instance before the fore foot leaves the ground in running. The triceps is a powerful combination of muscles. Its length, varies with the distance from the joint at the shoulder, being seventeen inches at its greatest and eight at the least distance. Its weight which is eight pounds, does not give a full conception of its power, for its action is nearly direct.

The *anconeus* is a small muscle attached to the capsular ligament of the elbow joint, and is inserted into the olecranon process of the

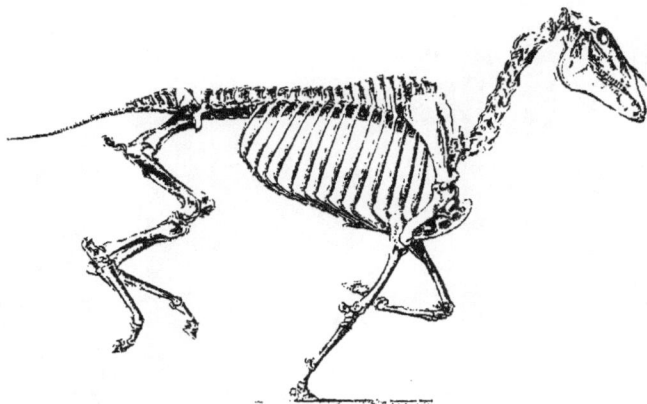

1

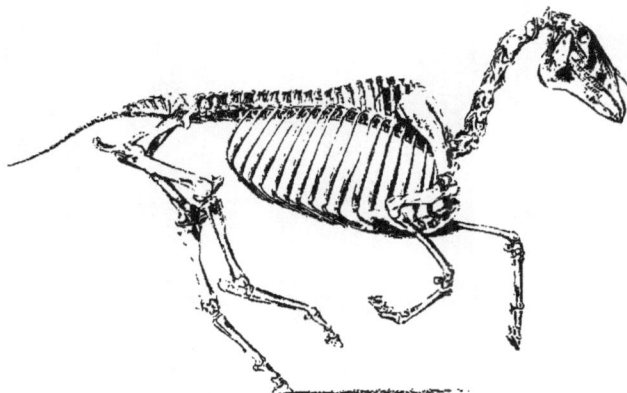

2

ulna, or point of the elbow. It contracts synchronously with the triceps, and its action is upon the capsular ligament to pull it out of the way and prevent its being pinched in the elbow joint as it becomes relaxed in the extension of the forearm.

The muscles of the forearm are, like those of the posterior extremity, simple and direct in their action in extending and flexing their levers, and, like those of the foot, their functions have been well studied and are well known. But the complex forces are the more difficult to understand the nearer we approach their sources, and have led to great diversity of opinion; the manner in which a movement was produced could not be explained for the reason that the motion itself was not understood. Now that the camera has rendered those motions easy of analysis, it is not difficult to show how they are produced.

The corresponding angles being reversed, the anatomical relations of the great flexors of the feet are changed. In the posterior limbs their tendons passed over the angles of the hock to be inserted into the bones of the feet. In the anterior extremities the corresponding tendons are enclosed in a sheath of the strongest possible construction, into the outer wall of which the pisiform bone is placed, to afford better protection to the tendons in the flexions of the knee joint, which is double, so that when the flexion of one is completed it is continued in the other, and greater flexion of the metacarpus upon the radius is effected than would be possible were the joint single. It will be noticed, on reference to the silhouettes, that the knee is never bent when the corresponding foot is on the ground. It plays its part in the rôle of a crutch consistently, but it performs a lively part in another character when relieved from the weight of that responsibility.

The tendons of the perforatus and perforans are utilized as ligaments as in the posterior extremities, but with some variations rendered necessary by the different conditions. From the posterior surface of the metacarpus, or cannon bone, below the knee, a ligament is thrown out to the perforans tendon to reinforce it, and other ligaments or tendinous connections are made to prevent extension of the joints beyond that of the standing position, by which the tension is taken from the flexor

muscles, and their tendons act as ligaments, their size being out of
all proportion to their use as tendons; and in the extreme extension
of the pastern the strain comes upon both tendons and the suspensory
ligament, and extension beyond that in the standing position is effected
only by the weight of the body, and at the expense of the elasticity of
all combined. While rupture of these tendons is of rare occurrence
under the strain thus put upon them, the sheaths through which they
glide above the pastern are not unfrequently torn transversely, giving
rise to inflammation and adhesions.

It is stated as a general proposition that the tendons are inextensible. This statement requires qualification. That they are so under
all ordinary uses *as tendons* must be admitted, but when they are serving as ligaments, in concert with the suspensory ligaments, they are
put upon a strain that muscle is incapable of applying or resisting. It
is therefore, as we have elsewhere shown, that powerful branches are
attached to the cannon bone to relieve the muscle of a strain it is not
capable of resisting. The organic tissue is tendon, but its use in extension of the foot beyond a certain point is that of a ligament to limit
extension; but as a ligament it does not absolutely arrest extension,
for it is elastic and allows of further extension after it is put upon the
stretch; or the extreme extension that takes place in the fetlock, to
allow the pastern to take a horizontal position, would not be possible. Further proof of this will be given after a quotation from Mr.
Percivall, which I shall give *in extenso*, for the reason that the information conveyed by it is very important, and few in America will have an
opportunity to consult his works, from their extreme rarity.

" The parts sprained are naturally supposed to be ' the sinews.' But
sinews or tendons, being both inelastic and (per physical force) inextensible, they themselves can neither be stretched nor strained so long as
they maintain their cohesion of substance. To discover, therefore, in
what part the sprain or lesion is likely to be situate, it will be advisable
to submit the leg in its normal state to anatomical examination.

" If we strip or dissect off the skin from the flexor tendons, we find
underneath, between them and the skin, a quantity of loose cellular
tissue, cutting away which we come to a close, or proper, tunic of the

same substance immediately enveloping the tendons. This under, or proper, covering, however, is fibrous as well as cellular in composition. For the space of a hand's breadth below the knee the glistening (tendinous) fibres may be seen crossing obliquely over the tendons, as they run from the annular ligament of the knee to be implanted into the external border of the cannon bone behind the external splint bone. This forms the sheath of the tendons. And when we slit it open we discover a cavity possessing a surface of a synovial nature ; and a sac, or *bursa*, thereby formed, which extends half-way down the leg, and is then closed. Through the bursa runs the perforans tendon, which may indeed be said to form a posterior boundary to it. The interval between the flexor tendons and the suspensory ligament, in their front, is likewise filled with interuniting cellular substance. This brief and imperfect anatomical sketch may serve to illustrate the nature of sprain. It will at once strike us that, although the tendons themselves are incapable of extension, and are too firm and strong in their texture to sustain hurt from any common accident, yet that they are surrounded and connected together, as well as to the parts contiguous to them, by a soft, delicate tissue which must, every time they are forcibly pulled or stretched, be extremely liable to stretch and lacerate ; and this, in fact, it is which in all ordinary cases constitutes the true and sole nature of 'sprain of the back sinews.' "

What is proved from the facts presented in the above quotation is that laceration of the sheath of the tendons could not take place except by the elongation of the tendon itself, and on that elasticity, or spring, of these tendons, in conjunction with that of the suspensory ligament proper, the mechanical action depends and in it their chief value consists.

The action of the anterior extremity as a unit in locomotion may now be studied. As the limb is thrown forward and in the act of taking the ground, it forms a straight line from the elbow to the heel ; the toe is raised, as in the posterior extremity, and contact is made with the heel. When the weight comes upon the foot the suspensory ligament is put upon the stretch by the reflexion of the pastern. The knee is kept in a straight position by the tension of the extensors, while the impulse is transmitted to the humerus at the angle of the elbow, too rapid flexion at that joint being prevented by the force of the triceps at the point of the elbow and the contraction of the great pectoral, which, acting on the shoulder, prevents the sudden flexion of that joint, at the same time that it forces the body over the limb; in which action it holds an analogous relation to the vastus of the posterior extremity, the application of mechanical power being utterly dissimilar, but the result in locomotion the same. The great dorsal aids in this office, though less efficiently. The superspinatus, acting from the scapula upon the upper end of the humerus over the shoulder joint, performs the same function for that joint, preventing its flexion too rapidly. The traction of the triceps upon the scapula is so great that it would be torn away from its position but for the counter action of the trachelo subscapularis, which transmits it to the cervical vertebra, as already explained. In this order the angles at the shoulder and elbow close while the fetlock joint is bent until the pastern is horizontal with the ground. In this action the limb is practically shortened, until from the position of the hypothenuse it becomes the perpendicular of a right-angled triangle, and during this change of position it has given uniform support to the centre of gravity without deviation of the direct line of its motion. During this time the levator anguli scapulæ has been contributing its force by acting on the short end of the lever, drawing it forward and adjusting the axis of the limb to its changing requirements. The passing of the body over the limb in a position perpendicular to the ground enables the limb in its character as an automaton to exert a propelling force as well as a sustaining one. It is necessary, however, that the support should be constant, as before; the angles must open as gradually as they had closed, and the fetlock joint must be as gradually straightened. Some changes take place in the action of the forces. The superspinatus, that had been yielding to allow of flexion, now contracts with greater force, and its labor is rendered easy by the continued traction of the two great propellers acting from the thorax. The branches of the triceps acting from the scapula relax altogether; that from the humerus by its continued contraction extends the forearm upon the humerus. This order continues to the close of the first

half of the stride, when it is in the power of the animal to give an
impulse to the movement, that settles effectually the question of the
power of the anterior extremity as a propeller. The proof will be
given hereafter, but the modus operandi is as follows: As the foot
is about to leave the ground, the angles of the limb being extended to
their utmost, the great pectoral, the great dorsal, and the great serratus,
by a vigorous and simultaneous effort, in conjunction with the spring
of the suspensory ligament and its reinforcing tendons, are capable of
deflecting the centre of gravity of the whole body of the horse, going
with a velocity of twenty miles an hour, four inches in a distance
of ten feet!

At the completion of the stride, the last impulse given by the
reaction of the suspensory ligament is like the spring of a bow, and
the flexor muscles regain control of their tendons, which had just been
serving as reinforcements to the suspensory ligament.

At this moment, the foot being off the ground, the knee bends
under the contraction of the perforatus and perforans muscles. The
superspinatus and all the muscles of the triceps are relaxed; the
flexors of the shoulder and of the arm contract; at the same time the
mastoido humeralis and superficial pectoral, acting on the shoulder,
carry the whole limb forward on its centre, the great pectoral and
great dorsal consenting. In this order they pass the perpendicular,
when the order is quickly reversed; in twice the speed used in the
retrograde movement, the foot is again in position to take the ground.
The extensors of the feet, after a rest of three fourths of a stride, again
straighten the knee, raise the toe, and the triceps is ready with all its
heads to take the shock. The sterno prescapularis pulls the slack out
of all the tissues connecting the shoulder with the trunk, that nothing
may be lost to effect the last line in extension.

The action of the anterior extremity in the three characters whose
parts it performs at the same time, we have endeavored to represent. In
these three characters it is a complicated machine. The object in its
construction was to enable the limb to support the body for the
greatest length of time, and to graduate that support so that it should
be uniform and constant, and that there should be no loss of momen-

tum or waste of power by correcting deflection of the line in which it
is intended the body shall move. This rendered necessary the use of
the legs alternately, so that while one should be performing these
functions the other should be moving in the reverse direction, to take
its place and permit as little loss of time as possible between the end
of the performance of one limb and the beginning of that of the next.

If we have comprehended the movements of a limb and the relative
value of the forces that produce them, the levers on which they act
and the relation of the limbs to each other, we ought to be able to
determine the mechanical elements of the qualities desired in a horse.
If speed is desired we must look for those mechanical conformations
of parts that determine speed, but this speed must be attained at the
expense of power. The anterior limbs must conform in their mechanical force to the posterior, and *vice versa*. It was observed of the posterior limbs that long full propellers (the vasti and glutei), low hip joint
set well back, so as to afford room for long femur and tibia, gave great
length of limb when extended, enabling it to support the weight of the
body and exert its propulsion for a longer time, at the same time the
power was more directly applied when the head of the bone was lower
down. So far as these principles can be applied to the anterior limbs
they hold true of them as of the posterior extremities.

It was observed by Bishop that all animals distinguished for great
speed have the angles of the bones most inclined to one another. But
while this mechanical arrangement gives great advantage for speed it
is a source of weakness in bearing burdens and hauling.

The requirements for the anterior extremities, to be in harmony
with the posterior ones, would be a long oblique scapula and long
humerus: these bones long, the angle formed by them would necessarily be less obtuse. The great pectoral and great dorsal are the
muscles that hold the mechanical relation to the anterior extremity
that the great gluteus and vastus do to the posterior; and to give equal
advantage to them the thorax should be long to give sufficient distance
between the ribs of origin and the insertion at the shoulder.

The application of propulsion to the anterior limb is unlike that to
the posterior, and as it is an advantage in the latter to have the heads

of the femurs low down to push more directly, in the former, on the
contrary, the centre of motion corresponding to the head of the
femur is not at the shoulder, but as high under the withers as possible, and the application of the propulsion as low as possible, as
was shown when describing the action of the great pectoral ; for,
the foot being the fixed point. the nearer the power is applied to it
the greater will be the velocity of the upper end of the extremity
acting as a crutch.

In the extreme of flexion and extension, as represented in leaping,
the muscles act at great disadvantage, as is illustrated by the difficulty
and slowness with which an animal rises from a recumbent posture.
They are positions incompatible with speed.

Whether the muscles act with the greatest energy at the earlier or
later stage of contraction has not been determined with certainty as
far as I know.　There is no doubt, however, that they act with the
greatest promptitude in response to the will when the limbs are
slightly flexed.　Boxers will instinctively put themselves in that position when in attitude for offence or defence.　Boys when about
to start for a race will relax their extensors to get a good send-off, and
they do not fully extend them again until the trial of speed is over.
So the horse in fast trotting "settles to his work," as it is technically
called.　In this expression it is intended to represent the idea that
the centres of motion are nearer the ground in order that the muscles
shall act to the best advantage. and that in propulsion the act shall be
most direct and longer sustained ; or, in other words, the points of
action and reaction are in a line forming a more acute angle with
the ground.

M. Weber asserted that the velocity in walking will be greater the
nearer the head of the femur is to the ground ; as this height increases the velocity decreases.　One sometimes arrives at a truth by
a very devious route, though he may have lost himself on the way.
He proved his position by the pendulum, which has been made to
demonstrate many a knotty proposition ; but while the leg of his
physical horse has swung three feet. our living horse has gone forty.
and his extremities have performed two complete revolutions.　The

11

speed of the horse does not depend upon the length of the limbs acting as pendulums, but upon the length and thickness of the locomotor muscles, the angles and lengths of the bony levers on which they act, the freedom of their articular ligaments, the correlation of all the mechanical parts, and much also on the nervous energy or will transmitted to the muscles, technically known as *courage.*

CHAPTER V.

THE attraction of gravity, or that force which is constantly drawing all bodies toward the centre of the earth, is a phenomenon so familiar to us that we fail to realize it at all times, and the consequences that would ensue were it to be for one moment suspended. Like the air we breathe, it is one of the necessary conditions of our existence, and the force with which it acts on all bodies is exactly measured by their weight; but this is the measure of that force in bodies in a state of rest or inertia. The instant that support is removed and the body yields to that force, there enters another element that must be taken into account, and that is *momentum*. While gravity is a constant quantity under similar conditions, momentum is a constantly varying one.

By yielding to the force of gravity an object does not escape from its power, neither is it reduced one grain in its influence at whatever rate the body falls. It is therefore an increasing quantity in a rapid ratio.* It is this force, which is constant and measured by the weight

* The formula for the determination of the distance which a body will fall in any given time is, $D = \frac{1}{2} g t^2$; in which D = distance; g, acceleration of gravity, or 32 feet; t, time in seconds. From this we learn that the distance which a body unsupported would fall in the first $\frac{1}{2}$ second would be 4 feet; in the first $\frac{1}{4}$ second, 7.68 inches; in the first $\frac{1}{8}$ second, 3 inches.

of the horse, that renders necessary the great development of the locomotive organs and columns of support. The power of resistance of
these organs must be equal to the attraction of gravity and counteract
it, and at the same time be in such excess as will afford the means of
propulsion in a horizontal direction. The influence of gravity is not
affected by motion in the body subject to it, at whatever rate it may
be moving. It may be projected into the air by a force greater than
that of gravity, but it does not escape from it in any degree. The
force that projected it was stronger than that of gravity at the time of
the impulse ; but the resistance of the air and the constant force of gravity would soon bring the motion to an end without the continuation
of that projectile force.

It is the result of this continuation of force in such directions as will
resist the attraction of gravity, and overcome resistance to a movement
in a horizontal direction, that we call locomotion.

There is another physical law to the effect that a body put in motion will continue in motion in the given direction until diverted by
another force from another direction. The force with which a body
moves above the surface of the ground is determined by multiplying
its weight by its velocity, and is called its momentum ; therefore the
force of gravity, represented by the weight, being constant in the same
body, the momentum will be as the velocity. If the body be represented by an iron ball weighing one thousand pounds, moving at the
rate of twenty-five miles an hour in a horizontal direction, it will
represent the momentum of a horse of equal weight at full speed.
To arrest it suddenly would be its destruction.

To continue its motion without diminution of velocity requires a
continuous application of force, and the greater the velocity the greater
is the necessity that the trajectory or line of motion should suffer no
deflection, for the force necessary to correct it increases with the
momentum.

From what has been said it follows that the only muscular power
required to keep a body in motion, at whatever speed, is that which is
necessary to resist the attraction of gravity and overcome resistance.
It is plain that, in order to maintain a uniform support of gravity, and

a continuous impulse in the direction of motion, the limbs must move, at whatever pace, in such manner as best to attain that end; that the more rapid the motion, the more uniform must be the support.

If the time occupied by a racing horse in going a mile be one minute and forty seconds, and the length of stride twenty-five feet, as represented of some horses, it would follow that he must be off the ground a full half-second at each bound, and according to the law of falling bodies he would, if he moved horizontally, during that time fall a distance of four feet. But in the gallop he is supposed to be moving by a succession of bounds in which he rises as far as he falls. This would give one fourth of a second as the time of descent equal to one foot of vertical fall to twelve and a half feet movement in a horizontal direction, and a consequent deflection of the centre of gravity to that extent.

We can imagine the effect that would be produced upon a railway car of any description if, when going at the same rate, it should pass over an obstruction that would produce such a deflection of the line of motion; and if instead of the railway car we substitute the horse, what but a broken neck could be expected? There is no suspensory ligament, or back tendon, or joint of the body that could be submitted to such a shock in a state of tension, and not go through bankruptcy

It is this deflection in the line of motion that constitutes the great obstacle to be overcome in all methods of locomotion. It is that which retards the progress of a ship in a rough sea; a certain amount of momentum is lost in every undulation, and power is spent in lifting the ship against the force of gravity that might in a smooth sea be spent in accelerating velocity. But a ship is an inanimate object acted upon by inanimate forces, and though speed is sacrificed, there is no exhaustion from waste of strength, as is sustained by all living beings in contending against the law of gravity, which requires a greater expenditure of force to arrest a body in its fall than is required to sustain it in a state of rest.

The most perfect method of quadrupedal locomotion, therefore, is that in which the greatest speed is attained with the least expenditure of vital force. This is found in those quadrupeds in which the devia-

tion of the line of motion from the horizontal is least. Pre-eminent
among these are the horse and hound, whose mode of progression is
the same. Though the deer and hare may have the advantage for
a short run, yet the method of progression by bounds, used by these
animals, sooner fatigues them, and in a fair field they will be run down
by the former from sheer exhaustion. To this subject we shall refer
again when we analyze the paces of the deer and dog.

In the two preceding chapters we have condensed the anatomy of
the locomotive organs into as small a space as possible, and at the
same time have, with the aid of accurate drawings, endeavored to
make the mechanical action of all the limbs *individually* so intelligible
that any one of ordinary information may comprehend them. We will
now proceed to show how these forces are co-ordinated in the production of the different paces when all the limbs are in action. Instead
of the mingled confusion of limbs and display of brute force, one may
see the most perfect order and regularity. In the slow movements
the limbs of the horse are without doubt much under the control of
the will; he may use his anterior ones to strike, paw the ground, and
in various ways show the control he has of different muscles in the
performance of their various functions; he may rear and kick, toss his
head and lower it to the ground, as in drinking, grazing, or sauntering;
but when speeding, whether from ambition or terror, all this trifling is
laid aside, the position of the head becomes fixed as a base of action
for the muscles of the neck and head, the detailed action of the various parts of the animal are lost in the complicated machine, and
the whole acts automatically, as the movements of the various parts
of a locomotive are lost in the combined action of the engine to
which they are subordinate.

The run is the perfect gait of the horse, for it is that which displays
most perfectly the play of all his locomotive organs, by which he at-
tains his greatest speed, and to which he owes his preservation in the
long struggle for existence through which he must have passed before
he came under the protecting care of man. It is the gait, therefore,
which best serves as a subject to study the law, or the theory of his
locomotion. To any one who has followed the anatomical analysis

of the last two chapters, the theory may have already outlined itself; but it is desirable that it should be made clear to all, and many of the anatomical facts demonstrated in the last chapters must be taken for granted in this by those who have not given the necessary attention to the anatomical descriptions. Perfect quadrupedal locomotion requires *uniform support to the centre of gravity and continuous propulsion by each extremity in turn.*

In order to avoid the abstract study of the co-ordination of the limbs in locomotion, figures are given to aid the mind in following the movements. They were executed by a process called photo-engraving, after drawings made with great care from a series of photographs, and represent twelve views of as many positions of a running horse. Three horizontal lines are drawn above the base at intervals of one hand, or four inches, as a guide to the eye in determining the elevation of the feet, and a fourth near the back to show the deviation from a horizontal line of the centre of gravity, which we will suppose to be under the saddle. These cuts are not introduced for their accuracy; they have been subjected to too much manipulation to lay claim to that precision of outline that will be found in the heliotypes and silhouettes in photo-lithography given in illustration of the paces.

FIG. 1.

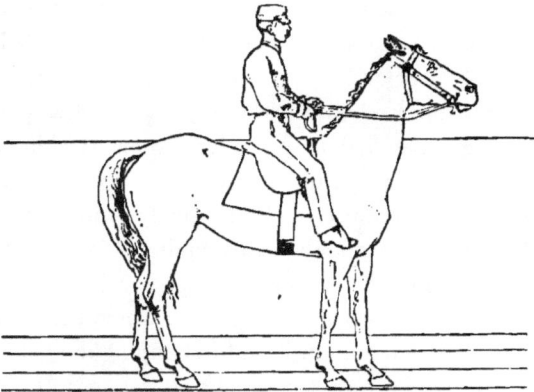

Fig. 1. gives the position of the animal in readiness to start. Its height is a little in excess of sixteen hands.

We have inserted two plates of the skeleton in the positions corresponding with Figs. 6 and 12 in order to enable the reader to understand their action in the various movements, and by reference to Plate II. he will be enabled to follow the descriptions in this chapter, and the action of the various muscles that produce them, as described in the previous chapters.

FIG. 2.

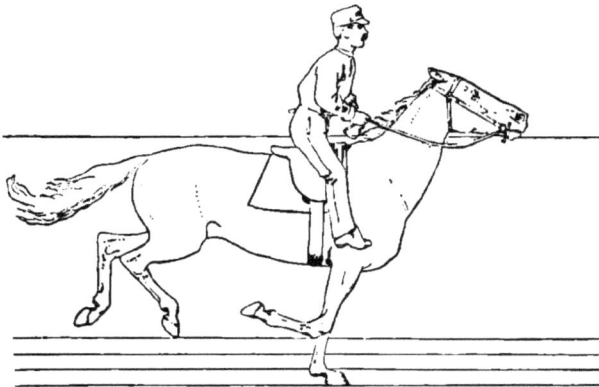

Fig. 2 represents the left fore foot upon the ground nearly under the centre of gravity; the centre of motion for the corresponding limb has passed in advance of the foot, and a line drawn through these two points would not be perpendicular to the surface of the ground; or, for brevity of expression, we will say he has passed the perpendicular. The limb is being elongated or extended by the straightening of the pastern joint and the joints at the elbow and shoulder; by these means the support given by the muscles is continued. In this position there is no muscular force exerted upon this limb below the knee. It was shown at page 76 how the "back tendons," while the limb is in this position, are converted into ligaments over which their muscles have for the instant surrendered control, and in conjunction with the suspensory ligament are supporting the weight of the body by their passive resistance. As the body advances by its own momentum and the continual action of the great pectoral and dorsal muscles, the pastern joint becomes nearly straight, and the pastern bone resumes its position under the metacarpus, or cannon bone, and while that joint is still supported at its convex posterior surface by these powerful ligaments ; and as the limit of extension is reached and the limb is at an angle with the ground of forty-five degrees, a vigorous concerted action of the propellers sends it forward and upward in the direction of the axis of the limb with a force so great as to close the space between the croup and the gauge line above it, and all his feet are in the air. If this force had been applied, as is popularly supposed, by the posterior extremity and behind the centre of gravity, the result would inevitably have been to pitch the animal headlong to the ground. The position of the fore leg just before it leaves the ground is best shown in Fig. 12.

FIG. 3.

Fig. 3 shows the feet all in the air. The foot which rested last upon the ground is now eleven inches above it and moving rapidly to the front. The interval of time between the photograph of the horse in position No. 2 and that of No. 3 was greater than that which passed between No. 3 and No. 4, owing to want of uniformity in the tension of the threads making the magnetic circuit. This defect was subsequently remedied ; and in the series of views illustrating the run, in which the gait was more thoroughly analyzed by a larger number of cameras, the intervals are very regular.

FIG. 4.

In No. 4 the upward impulse given by the fore leg may be supposed
to have reached its limit, the croup having passed above the gauge-
line marking the elevation, and the feet are all gathered under the
trunk more than a foot above the ground.   There is now an opportunity for the animal to change their order; for, as has been stated,
he cannot change the order of his feet when one of them is on the
ground, and he is going at a rapid rate, without subjecting himself to
a fall.   Although the distance passed over from the time the last foot
left the ground until the next one reaches it is only equal to the
interval of any of the feet, and the time that has elapsed while this
has been taking place does not exceed perhaps a fifth part of a
second, still it is sufficient to enable the animal to choose whether
he shall lead with one foot or the other in the next stride.   But the
time, short as it is, is sufficient to cause a descent of four inches, and
the momentum acquired makes the contact with the earth a much
more serious matter than in any other portion of the stride.   To make
it with either of the anterior extremities, or both, as is popularly believed to be the case, would seriously check the momentum, if it did
not result disastrously to the unlucky member.   But this has been
shown over and over, by numerous observations at Palo Alto, never to
take place in running.   The first check to the descent of the centre

of gravity is given by one of the hind legs, and by that one which is diagonal to the fore leg leaving the ground last; but, to reach it in such manner as to prevent a catastrophe, it must be planted beneath the centre of gravity or in advance of it, and then, in order to prevent the anterior part of the body from falling forward, it is necessary that all the available force should be brought to bear upon the right hind leg as a lever with its fixed point upon the ground. How this is effected, and by what muscles, is shown at page 58.

FIG. 5.

This situation is shown at Fig. 5. The picture was taken almost at the instant of contact by the right hind foot with the ground. The anterior portion of the body is arrested in its downward course, not by its own limbs, but by the contraction of all the muscles forming the external periphery of his body, from the neck to the flexors of the foot; by which combination of forces the whole body forward of the head of the femur is not only arrested in its downward course, but lifted, while the momentum in a horizontal direction is maintained chiefly by the contraction of the vastus shortening the distance between the lower extremity of the femur and the spines of the sacrum, pushing the pelvis forward from its fossa behind the head of the femur. The body is propelled forward with a part of the same power that lifts it. In this manner it does not check the momentum ac-

quired, as would have been the case had one of the fore legs been
thrust forward to the ground, and the danger of stumbling is averted.

FIG. 6.

As progression continues, the limbs are all taking positions in the
order they will be required to perform their functions.   The right leg
in Fig. 6 is passing the perpendicular; the pastern is bent to the
ground to shorten the limb, and the left hind foot is descending to
repeat the same action, when the right, from the advancing position of
the body, will be unable to continue its support, and the right or diagonal fore leg is straightening to take its turn after the left hind one.

FIG. 7.

In Fig. 7 the left hind foot is supporting the weight, and already the centre of gravity has passed over it, while the right, relieved from that duty, is exercising its function as a propeller, and the right fore leg is reaching forward to take the ground as far as possible in advance; the foot is extended to bring the heel, with its elastic cushion, first in contact. The left fore leg is straightening to take its place in the order of succession.

FIG. 8.

The right hind leg has given its final propulsive impulse in Fig. 8, and the tensor vaginæ femoris, the iliacus, sartorius, and superficial gluteus are in the act of flexing the stifle and advancing the leg to a new position. The left hind leg has passed the perpendicular, and is no longer in a position to give much aid as a supporter to the centre of gravity; but the right fore foot has reached the ground, and takes its position as a supporter of the weight of the body, dividing the burden with the left hind leg still upon the ground.

In the last chapter it was shown how the limb is thrown forward into this position, and how the shock of contact is transmitted, through the straightened extremity, to the humerus and scapula; and unnecessary flexion of the elbow and shoulder joints is prevented by the triceps of the arm and superspinatus muscles, while it continues to give uniform support to the body, at the same time that it shortens by the

gradual flexion of those joints and the bending of the fetlock as the
body passes over its point of support.

The danger to be apprehended in the use of the fore leg to arrest
the downward movement when the body was falling, as in Fig. 5, does
not exist in this position, as the momentum of gravity has been arrested by the posterior extremities, and the centre of gravity has
reached its lowest point, while the weight is divided with the hind leg.
Propulsion is now going on in the fore leg through the great pectoral and dorsal, and in the hind leg through its propellers proper.

FIG. 9.

The right fore leg in Fig. 9 is now taking the entire weight of the
body, is nearly perpendicular, is correspondingly shortened, and its
fellow is extended forward in the position to take its turn. The left
hind foot is clear of the ground, and the right has been elevated by the
action of the semitendinosus slightly flexing the stifle. The settling of
the body has not varied much from the position seen in the last figure.
Propulsion by the great pectoral and dorsal on the fore leg, which is
also bearing alone the full weight of the body, is most energetic, and
to the best advantage.

In the interval that has passed between the position of Figs. 9 and
10 the left fore foot has descended twelve inches, is only four inches
from the ground, and must be considered as in the position in which

FIG. 10.

the heel makes contact with it. The right leg is elongating, and
making propulsion by so doing, in its function as an automaton and
also as a passive tool, by the great dorsal and pectoral making traction
from the flanks upon the shoulder. The support it is giving to the
weight of the body is shown by the narrowing of the space between it
and the gauge line. This action is not yet complete. Both the hind
feet are nearly equally elevated, but the right leg is more flexed. The
increased flexion of the stifle renders necessary the flexion of the hock-

FIG. 11.

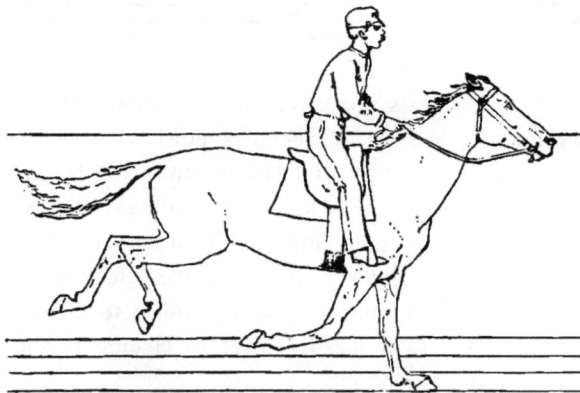

joint, for, as already shown, they act together automatically. The
flexors proper of the thigh are now making their force felt with that of
the semitendinosus, and the stifle is being drawn forward. The flexors
of the foot, being propelling muscles, are inactive, and the semiflexed
position of the joints of the feet is owing to the reaction of the suspensory ligaments.

In Fig. 11 the right fore foot is clear of the ground, and the left is
in a position corresponding with that of the right in Fig. 9, but the
right is not in a corresponding position with the left in the same
figure. The right hind foot is preparing to take its place in the order
of succession, to be followed by the other hind foot in its turn.

FIG. 12.

In Fig. 12 the stride is completed. On comparison with Fig. 2, it
will be seen that the body is advanced somewhat beyond that in the
latter, and it is again about to leave the ground with the left fore foot,
and the energy of the propelling forces have already sent the body up
nearly to the horizontal gauge-line. This order in the movement is
continued until the animal feels fatigue in the left fore leg from the
continued use of it in giving the final impulse to clear the ground,
when, as before said, it is in his power to change it and leave the
ground with the other fore leg.

It seems, at first thought, from the manner in which the labor is thrown from the leading fore foot to the diagonal hind one, during which the body has no support from either, that the theory of constant support and continuous propulsion does not hold true, — that a perfect machine should require no such hiatus. If the machine had been constructed of inorganic and inanimate material, incapable of fatigue, it could have been so arranged that the hiatus would not have been necessary; but the Creator did not, if he could, build an animate machine that would not tire. The animal is shown in the above figures as moving his feet in the same order, and, but for some arrangement that would permit of a change, fatigue would be inevitable; but that change would not be possible until all the feet are clear of the ground. If the attempt should be made while one foot is on the ground, the result would be called a misstep and a fall. The opportunity is afforded when the extraordinary propulsive force, given by the fore leg that leaves the ground last, projects the body upward, giving a time equal to one fifth of a stride for the hind foot of the same side to take the place of the one that would have followed had the same order continued. The coincidence of the act of changing the order of the feet with the exposure of the negative plate in the camera, which is too short a time for computation, must be very rare, and has not been observed in any of the many pictures taken at Palo Alto; but the change is felt by the driver when the feet again take the ground, and it is said to have been discovered in the changed order of the footprints.

While the run requires that each limb in turn should act as propeller and supporter in regular order, it cannot be executed at a low rate of speed, for the base of support is confined to one foot, and it must be rapidly adjusted to the changes of the position of the centre of gravity, for the same reason that a boy on stilts requires to be continually in motion. If the horse's speed is diminished to a great degree, he will change to a canter, which is a pace modified from the run, as may be seen by reference to Plate XXVII., or, what is more usual, to a trot, in which he uses two diagonal feet as bases of support.

The order and action of the limbs are uniform in all the numerous trials of running horses photographed at Palo Alto; so that it may be considered that they are in conformity with a law. Five series of the photographs are given, of horses representing different rates. The first is that of "Mohammed," whose stride is 15 ft. 9 in.; and the last, that of "Florence Anderson," with a stride of 20 ft. 6 in. All of them were going at a moderate rate. The numbers, and corresponding lines on the ground, indicate spaces of one foot; and as the photographs were taken in succession at the same intervals, they will be understood to show the position of the limbs at each interval of one foot. The position of the camera is indicated by the direct line on the ground, and is that of the observer. In the last series the gauge line shows how little the centre of gravity is deflected in its trajectory from a direct line, and this line will be found to vary least when the speed is the greatest. By the aid of these lines and their figures the reader may be able to measure the strides and parts of strides, and determine their respective intervals.

It has been observed that there is not perfect regularity in the line of the footprints of a running horse, especially if the ground is uneven. This is owing to the variations of the centre of gravity, which compel the corresponding variations of the positions of the small base which supports it; through an instinct of the same kind which we recognize in ourselves, and make use of when we fail to give proper attention to the ground on which we are walking and govern the movements of our feet accordingly. This we do not always do, and the effect is to cause us to stagger even when we are sober. This instinct must be regarded as existing in a higher degree in a horse, as his locomotive apparatus is more complicated and of a higher order than that of man.

If the reader is interested in knowing how far this law of the mechanism of running holds, he may follow it in the succeeding plates. He will see the same movement in the greyhound, Plates XVIII., XIX., and in Plate XX. two hounds running at unequal rates of speed. It is the same in the ox running, Plate XXI., and has been proved true of the goat, and will be found to hold true of all those quadrupeds whose four limbs are of like proportions.

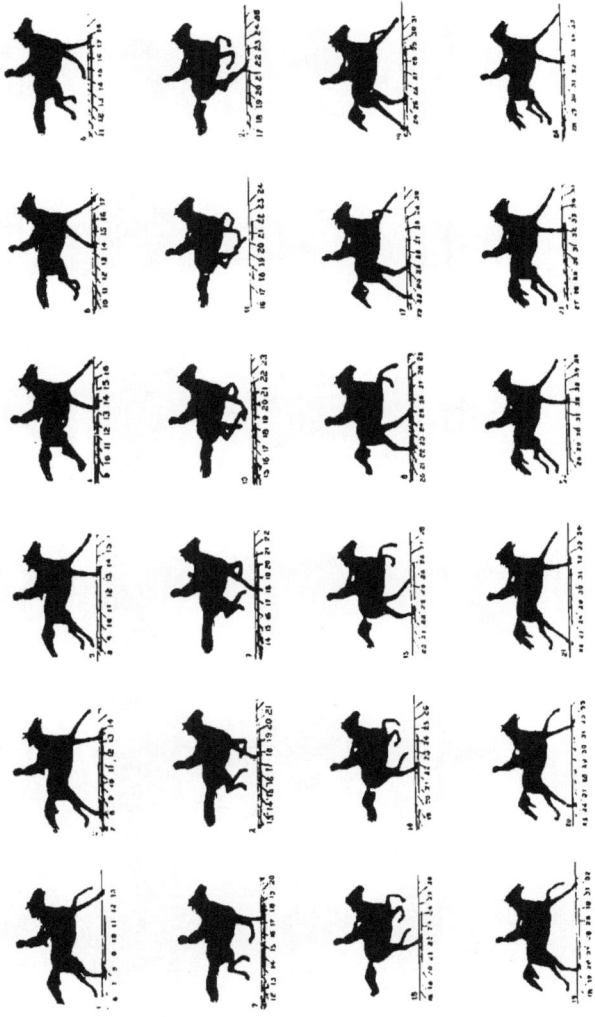

PLATE XV

FLORENCE A. RUNNING, STRIDE 19 FT. 9 IN.

PLATE XVI.

PHRYNE I. RUNNING STRIDE 6 to 9 ft.

FLORENCE A. - IRUNNING. (Males 29 and 31.)

PLATE XVII

HOUND RUNNING

PLATE XIX

20 21 22 23

25 26 27 28 29

29 30 31 32 33

18 19 20 21 22

24 25 26 27 28

28 29 30 31 32

17 18 19 20 21

23 24 25 26 20

27 28 29 30 31

16 17 18 19 20

22 23 24 25

26 27 28 29 30

PLATE XX

TWO HOUNDS RUNNING AT DIFFERING RATES OF SPEED

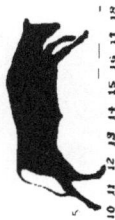

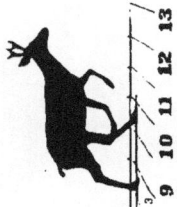

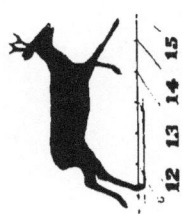

There is another class of quadrupeds, whose posterior extremities are developed to a much greater degree than the anterior ones, and their mode of progression varies more or less from the theory of motion as given in the preceding pages. The only one of this class of whose stride we are able to present an analysis is the deer. In this animal the same order of succession of the feet may be observed; but, at the moment when it might be expected that he would, as in the horse, rise from one of the fore legs, owing to its feebleness he fails to do so, but the hind legs are thrust forward to the ground, one of them in advance of the fore foot and the centre of gravity, and, while one of the fore legs still supports the weight of the anterior part of the body, by the combined action of the posterior extremities he projects himself by a bound, and alights, not upon one of the hind feet, but upon one of the anterior extremities; but the action of the limbs in pairs is not synchronous, one of them being a pace in advance of the other, to distribute the shock. This mode of progression the deer is able to perform by reason of the length and angles of the bones of his limbs and the lightness of his body, and it is adapted to the nature of the ground among the hills where he finds his only safety, and where the horse would be at as great a disadvantage as the deer is upon the plain. For reasons already stated, it is not possible to sustain this gait for a long time from the exhaustion which it produces. The action of the horse in leaping will be reserved for another chapter.

From the knowledge we have gained, by the use of instantaneous photography, as to the action of the feet in running, the answer to the questions propounded by William Percivall thirty years ago is obvious: "What is the reason why the flexor tendons fail so much more frequently than others? Another, Why those of the fore limb should fail rather than the flexor tendons of the hind leg." *

In the following quotation he gives the answer to his own question, in accordance with the heretofore accepted theory of the run. "I have more than once had occasion to direct attention to the important functions performed by the hind limbs in the acts of progression, and to contrast these with the comparatively light duties of the fore limbs.

* Hippopathology, Vol. IV. p. 346.

While the former, like a pair of oars at work in a boat, are plying for-
wards and backwards, forcing the body onward, the latter, more like
stilts, are employed in sustaining the propelling parts, lest the body fall
forward to the ground.   I have likewise afore observed, that two such
different functions necessarily distress different parts of the limbs, —
the hock being the part most exerted in the hind, the feet and legs the
parts most tried in the fore limbs.   What distresses the sinews of the
fore limbs so much is the extreme distension, almost preternatural, to
which these legs are put in hard galloping and leaping every time the
weight of the body descends upon them, at a moment when they are
stretched out to their utmost, as they must be, to receive it; and it is
to this identical position of the limb, whenever any weight or force
of extraordinary amount, or in any sudden or unexpected manner,
descends upon it, that *strain* or *sprain* is produced." *

It is now perfectly clear that it is in their action as propellers
that the flexors of the fore leg become injured in their tendons, and
in the position shown in Fig. 12; though there is little doubt that
if the weight of the rider were not superimposed to that of the horse,
this accident would rarely happen.   Sprains involving the ligaments
of the joints may occur at any time when the foot from any cause is
not set squarely upon the ground, but this is an accident of quite a
different nature from that to which reference is made above.

We have designated the pace under consideration as the run, —
the pace used in racing, or the fastest known to the horse and other
domestic animals.   What, then, is the gallop?   If we are to be confined to the definition of the gallop given in the dictionaries, and
generally accepted by all writers on the horse, we are forced to the
conclusion that there is no such pace, that it is a fiction.   Webster
defines the gallop as "a mode of progression by quadrupeds, particularly by a horse, by lifting alternately the fore feet and the hind
feet together, in successive leaps or bounds"; and Worcester, "to
move forward as a horse by such leaps that the hind legs rise before
the fore legs quite reach the ground."   That the pace which we call
the run is not such as will bear the definition given is very clear.

* Hippopathology, Vol. IV. p. 346.

The camera has, under the direction of Mr. Stanford, been made to analyze all the paces, and none has been discovered that answers to it ; yet it is to this pace that the term "gallop" has been always applied.

When, three or four years ago, the results of Mr. Stanford's experiments with twelve cameras were distributed in art circles, the photographs sent met everywhere with surprise and incredulity, and in some quarters with ridicule and burlesque. Such result ought to have been expected. They were not understood, and the revelation was so antagonistic to all received opinions from the earliest times, that one could not help but laugh ; and that they do not understand them now does not surprise us, for hippo-anatomy has been always taught under the light of a false hypothesis. When we consider how little the simple movements of the trot were understood by the most learned of the teachers of animal motion, is it a matter of wonder that the complicated mechanism of the run had been kept so profound a secret in the open face of day from time immemorial ?

A revelation so startling as that made by the camera carried results too far-reaching and revolutionary to be at once accepted, though it came direct from heaven. There is too much capital invested in works of art all over the civilized world to permit the innovation without protest, and ridicule is the cheapest argument that can be employed in controversy, for it does not require truth for its foundation, and but a low order of talent for its display.

All artists know the value of the horse as a *chef d'œuvre*, and he is made, next to the human figure, the first subject in elementary studies in art ; but from what source have been derived all the models of horses in motion? Who does not tire in looking over the monotonous repetition of outstretched legs, as if their bearers had been shot from a cross-bow, and were moving at a mark without any agency of their own, and when the slightest variation of that inflexible form would spoil the pose and ruin a picture. We are told that the object is to represent *action*. Would not that object be more readily attained if some position were represented that is known to be true, instead of one that is proved to be impossible ? " But art must represent things as they seem," says an objector. Those who think they see a horse in

the positions given in the conventional way have their conceptions
formed by a false hypothesis; those who are initiated into the true
theory of the movement experience no difficulty in perceiving the
movements of the limbs in precisely the manner represented in the
plates here given, and wonder they never saw them so before. One
will recognize this movement more readily in a dog running than in
a horse.

Some of these positions seem grotesque, but for no other reason
than because the action is not understood. When it is so, they will
appear as necessary progressive stages in a never varying series of
movements, the result of which is locomotion, and it will appear that
it cannot be performed without them; the eye that understands them
can never be deceived, and the slightest deviation from the law of their
co-ordination will instantly be detected in a *silhouette* as surely as the
animal would be to suffer the consequences of a misstep.

Quadrupeds will be recognized as being possessed of locomotive
machinery, self-moving, with all the parts acting in perfect harmony,
and not passive projectiles. If Art is the interpreter of nature, as is
claimed, she is false to her mission when she wilfully persists in per-
petuating a falsehood. But in this case she cannot if she would.
Once attention is called to the true theory of quadrupedal motion,
the truth of each one of these positions, and the interpretation of them
in relation to progression, is so quickly recognized, while the error of
the old theory of the gallop becomes so manifest, that artists will no
more be able to claim that they represent nature as she seems, when
they depict a horse in full run in the conventional manner, or the
mythical gallop.

Plates XXV. and XXVI. represent sketches taken from elementary drawing-books manufactured in London and Berlin and used in
the schools. They are heliotyped, on a reduced scale, in order that
there shall be no suspicion of inaccuracy in the copies.

After what has been said, comment is unnecessary; but I would
ask, if animal motion is to be always taught to follow such severely
false models, wherein is it better teaching than that of the priests of
Osiris, with whom all forms were stereotyped for thousands of years,

and the last stages of their art were worse than the first? And here
I would diverge still farther from the path I had marked out for
myself, to protest against the soundness of that dogma that art should
represent things as they *seem*.   I will not enter upon a discussion of
the general proposition, it would carry me far beyond the special
subject of this essay; but I will limit myself to the consideration
of the proposition as it is applied to the representations of the movements of the limbs of the horse in motion. I am often told that
we do not represent the spokes of a carriage-wheel in motion as distinct spokes, but they are made to run together as they really appear
to the eye, where new images are made upon the retina before the
first are lost.   Lightning is represented as a zigzag line, when in
reality it is a spark; but this spark moves with such inconceivable
rapidity that it is quite impossible to calculate the time of its passage.
The *track* of the electric spark is unquestionable, and when that is
represented there is no untruth: it would be impossible to represent
lightning in any other way.   Lightning is not only the spark, but
the track it describes in the sky also. To represent them thus is
to represent the actual truth, and so it expresses action, and no error
is inculcated.   But does one ever represent the horse's legs in that
manner to express their action?   Why not? If it cannot be done,
why assume a false and hackneyed position that cannot be true?
Why must every equestrian statue in Europe follow the model of the
Antique Balbi in the Neapolitan Museum, with the bones of the fore
leg flexed at right angles, and the other three feet upon the ground?
No such position was ever true, nor can it seem to be so to any one
who gets his impressions from nature.

The theory of the run and the mechanism of the locomotive
apparatus of the horse will soon be common property among admirers
of the animal; and when that knowledge becomes general, all the
famous paintings in which he is a prominent figure in the "gallop"
will be relegated to the museums as examples of old masters, to
illustrate the progressive stages in the development of art.

If the theory of the run is understood, the action in the canter will
present no difficulty; for the theory is the same in both, and the varia-

tions in the order of movement of the feet are changes rendered necessary by the low rate of speed in the latter. In Fig. 1, Plate XXVII., the cantering horse is seen in the act of leaving the ground with one fore foot as in the run, and his feet are clear of the ground only for the distance of two feet and five inches, as indicated by the lines on the ground, when the diagonal hind foot comes to the support of gravity, not under its centre, as in the run, but behind it (see Fig. 4), and therefore cannot prevent the body from falling forward. In order to prevent this result, it is necessary that one of the fore feet should support it, and it is always that fore foot which is diagonal to the hind one that is upon the ground. The other hind foot follows at the usual distance from its fellow. He has now, through three Figures, three feet upon the ground, as in the walk, after which the order of the run is resumed. Fig. 11 nearly corresponds to Fig. 1; the difference observed is owing to a want of correspondence in the time of exposure of the sensitive plate of the camera. In Fig. 1 the fore leg has given its quick thrust, and the knee is slightly bent as it is about to leave the ground, while in Fig. 11 it is still acting as supporter and propeller.

It is clear that if the animal moved with more will and greater speed, planting his hind foot farther forward in support of the centre of gravity, there would have been no necessity for the fore leg to have performed that office; and the pace would not have differed from the run. The length of time during which three feet support the body gives time for the rider to settle in the saddle, and causes that easy cradle-motion which makes it a favorite gait with ladies.

CHAPTER VI.

THE leap cannot be properly considered as a pace; although it is a mode of progression, it is not a continuous one. Before any quadruped will venture to undertake it, he must have acquired a considerable degree of experience in locomotion, and that confidence in the use of his limbs which experience only can give. That it is naturally acquired, there cannot be a doubt, as it is necessary to all quadrupeds in a wild state to enable them to overcome obstructions otherwise insurmountable. The weight of the horse's body, however, renders it necessary for him to reduce his speed, and with it his momentum, before he can safely attempt it, even when the obstruction is of moderate height. It is a feat in which he is excelled by most quadrupeds, all the quadrumana, and even by man himself.

From the mode of action of the various parts of the locomotive machinery, as shown in Chapters IV. and V., the reader will experience little difficulty in understanding what takes place in the leap. The action is so full of interest that we have given a number of illustrations to enable the reader to observe the many different phases the leap

presents. All the plates show the horses approaching the barrier at a
run ; but no sooner is it observed than they begin to shorten their steps
and apparently measure its distance. In Plate XXVIII. the hurdle is
placed at an elevation of three feet six inches, and the horse betrays
his anxiety by shortening his paces, and advancing with both hind feet
nearly simultaneously and alternately with one fore foot, or what is
called prancing, until he has approached the barrier sufficiently near
to satisfy himself that he can surmount it, when he plants his hind feet
well under the centre of gravity, and instantly the fore leg resting upon
the ground gives the thrust explained in Chapter IV., by which the
anterior portion of the body is raised, and the action is immediately
followed by all the muscles of the haunch, which project the body to
the required height. The anxiety and want of confidence of the
animal are betrayed by the nearness of his approach to the obstacle
and the arrest of his momentum before he ventures the leap. By
the arrest of his momentum he has diminished the danger of injury
to the back tendons on reaching the ground again.

In Plate XXIX. we see the same horse under somewhat altered
conditions. The hurdle is six inches lower, and he advances with
increased confidence, leaving the ground eleven feet from it.

The relations of the levers, or passive parts of the machine, in the act
of leaving the ground in leaping, are shown in Plate XXXV. Fig. 1,
where the positions of the posterior extremity are the extreme of
extension.

The next plate represents a full series of views by twenty-four cameras, by means of which the movements in leaping are carried four feet
farther. The posterior extremities from the extreme of extension, on
leaving the ground, pass to the opposite extreme of flexion as they pass
the barrier, and both the posterior and anterior limbs, as they pass successively in pairs, are so nearly in unison that they seem in the silhouette to coincide. Plate XXXII. shows the same horse as seen in the
last preceding plate, after he has passed the hurdle and is nearing the
ground. The anterior extremities, that coincided in passing the hurdle,
are now separating in order that they shall not make contact with the
ground at the same time. One of the fore legs is extended as in the

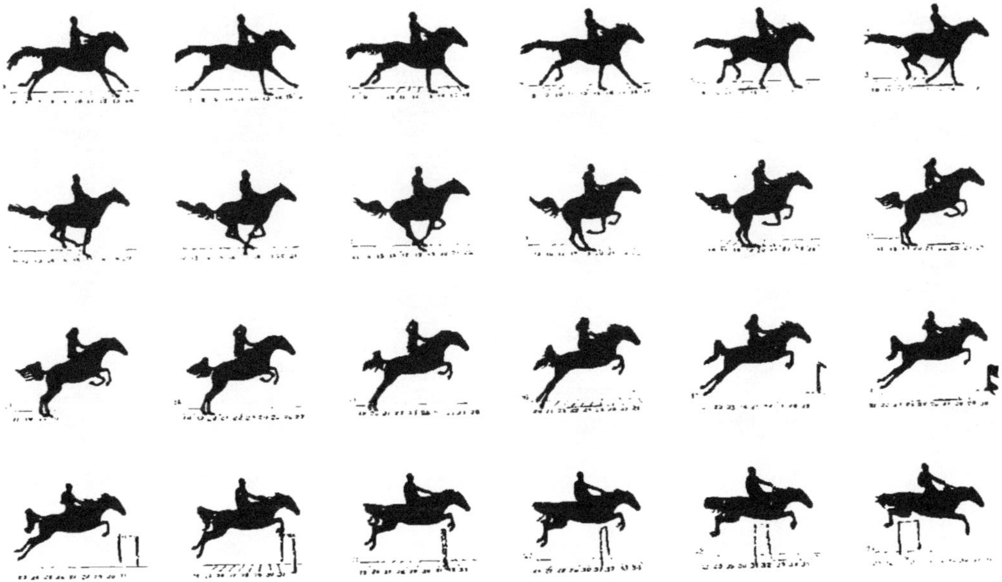

FIGURE

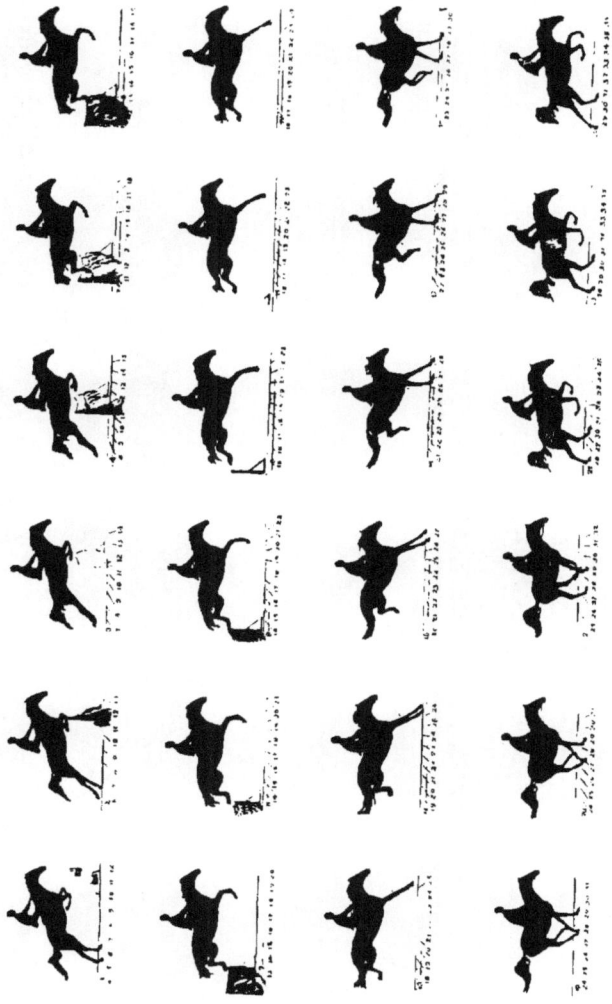

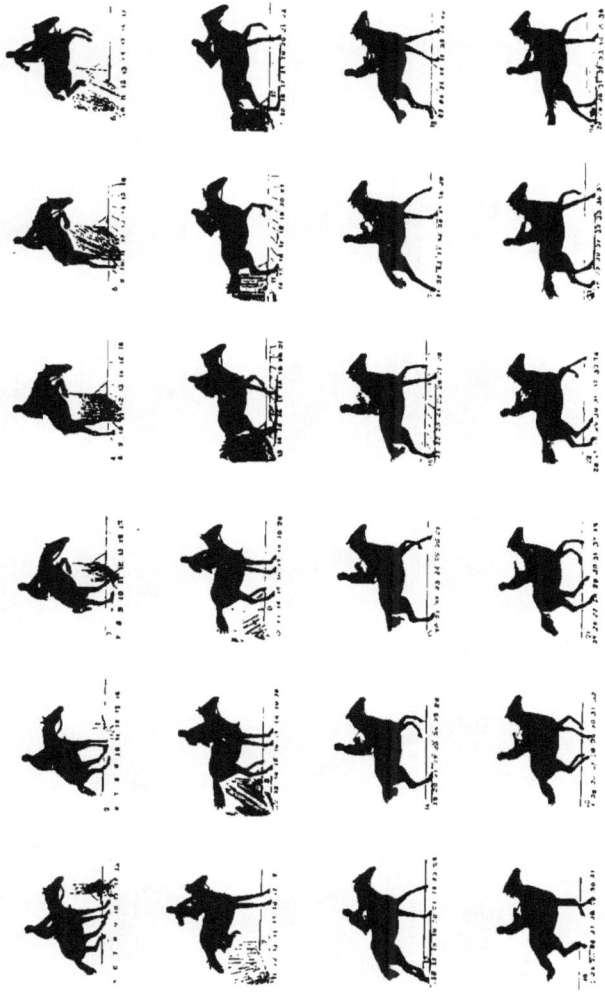

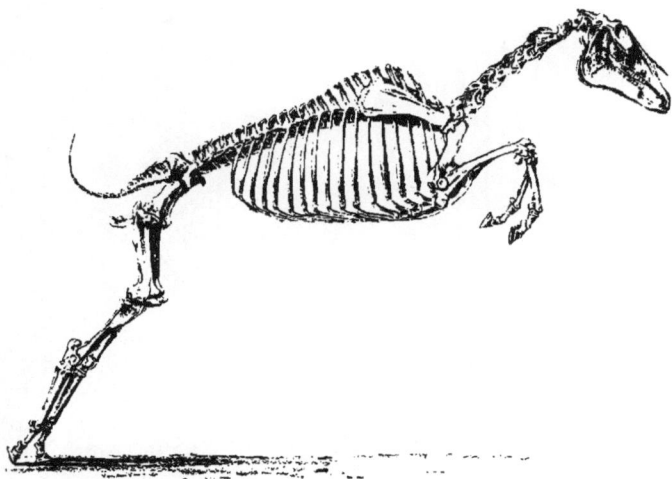

1

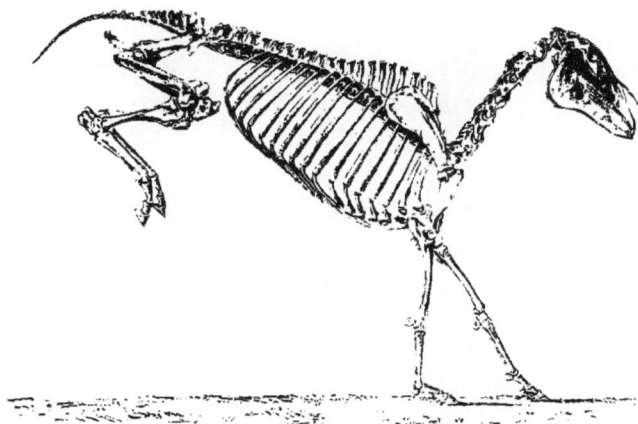

run, to check the force of the descent, which, from the loss of horizontal momentum, has little more than the momentum of gravity to deal with. This is the moment of extreme danger to the pastern joint and flexor tendons; but before these parts are put to the extreme test the other fore leg comes to the relief of its fellow, and immediately after the posterior extremities, one after the other, are planted under the centre of gravity, and by their great lifting force relieve the anterior extremities, and all the limbs are free to act their various parts in the run, which is not fairly resumed in this series, the velocity at no time being sufficient to enable the animal to clear the ground. The action after the leap may be still further traced in Plate XXXIII. where the run is not yet fully resumed, the speed being only equal to the disposition of the limbs as in the canter, the order being the same as in the run. The last illustration of the leap that we offer is very curious. The horse was very reluctant to perform the leap required of him, and came to a standstill immediately in front of the hurdle, and only after great urging did he attempt to surmount it. The action of the locomotive organs is shown to be the same in this as in the other series representing the leap, only with less courage manifested; and there is little danger in its execution. As the horse lost his horizontal momentum before leaping, so he had none when he reached the ground on his descent. It should be observed that in all these series the intervals between the successive pictures are those of *space* and not of *time*, as the horse makes his own pictures in a manner that will be fully explained in the Appendix. The intervals in all the series of twenty-four pictures represent distances of one foot in a horizontal direction. Fig. 2, Plate XXXV., shows the position of the skeleton as the animal meets the contact with the earth. By means of these skeleton views the reader is enabled to build up the entire locomotive apparatus with the aid of the anatomical plates, and satisfy himself as to the forces that are employed in producing the movements.

On reference to Plates XXIII. and XXIV., it will be perceived how great is the correspondence in the action of the deer in bounding and the horse in leaping. In both the action on leaving the ground is the same. When the hind feet are upon the ground, well under the centre of gravity, the spring of one fore leg lifts the anterior half of the body, and, the action of the posterior extremities immediately following, the whole body is projected into the air; but, the deer being in more rapid motion, his feet take the ground at longer intervals and more regular order, and so diminish the danger of stumbling, as well as distribute the shock of contact and equalize the support of the weight of the body.

When the horse reduces his speed in running so that he can no longer maintain his balance upon one foot, he will usually drop into a *trot*, which is a gait having two feet as bases of support instead of one. The theory of the trot is the same as that of the walk, but adapted to a higher rate of speed. It differs from a walk in that the latter has always two feet upon the ground, while in the trot there is always a space of time, of greater or less amount, in which all the feet are off the ground. Other differences will be noticed when we come to analyze the walk. They correspond in that the weight of the body is borne by the diagonal extremities alternately, and in the general co-relation of the limbs in their mechanical action. The action in the trot is the more vigorous as the distance in which the body moves unsupported is increased. The definition of the word *step* in its general use is somewhat ambiguous. It is often used synonymously with *stride*. In the step of both quadrupeds and bipeds it is understood to mean the distance spanned by the two feet both resting on the ground. This will vary with the muscular energy, but is limited by anatomical proportions.

The *stride* is here used to signify the distance passed over by one foot from the time it leaves the ground until it reaches it again, measured to corresponding points, and is equal to two steps; but in the trot this definition will not hold good, for there must be added a certain distance, differing according to speed, in which neither of the feet will be on the ground. If a represents the step in the walk, and x the distance passed over by the foot after the other foot is raised, the step in the trot would be expressed by $a + x$, and when a is constant, the step will vary as x. In the run, there being four steps, and an interval when all the feet are off the ground equal to a step, the stride would be expressed by $5\,a$. The stride is divided in the trot into two periods by the alternate feet, so that in the trot the horse is twice clear of the ground in each stride. The step being supposed to be a constant quantity in the fast trot, the stride can be extended only by increasing the space which the body passes over with its centre of gravity unsupported. While in the slow or jog trot this distance is small, in the flying trot it exceeds that in which the body is supported, and hence arises the great difficulty in attaining a high rate of speed. As was stated in the preceding chapter, the law of falling bodies increases the difficulty in locomotion in the ratio of the square of the time in which the body is so unsupported. It becomes a question of power of resistance, or strength of the parts on which that resistance depends. On the other hand, the strength of the parts, as the joints, bones, ligaments, and tendons, involves increase of weight, which is incompatible with rapidity of movement, without a corresponding increase of power of the muscles and weight of the body to be borne, so that the limits of speed attainable in a trot are reached more rapidly than in a run, in which the limit is to be found in the measure of activity.

In the run the stride is divided into five parts, instead of two, as in the trot, each limb taking its turn as supporter and propeller, with a scarcely appreciable interval between, and an interval between the last fore leg and the first hind one representing a fifth of the whole stride. Each limb, therefore, works one fifth of each stride and rests the other four fifths. The longest stride given of the run, in the examples furnished, is that of "Florence A." (Plate XVII.), where it is given at twenty feet six inches, or a little more than four feet as the portion assigned to each limb. It will be observed that in the trotting horse (Plate XXXVI.), whose stride is given at eighteen feet three inches, the time of support by two limbs is about the same, while the time in which there is no support given is greater, and divided into two intervals. So in Plate XL. the gravity is supported about half the time by two limbs, and the other half by none, alternating every four feet. Notwithstanding the wonderful mechanical provision in each of the four limbs to secure uniform support and propulsion while the feet rest upon the ground, the instant that the body ceases to be supported it becomes subject to the law of the descent of falling bodies and all its consequences, as mentioned in the last chapter, and the greater the speed of the animal the more serious the possible consequences ; and though no small advantage is gained by relieving the horse of the weight of the rider, and placing it upon a sulky, it is the cause of serious damage to the finest stock. It is no small accomplishment in a horse, however thoroughbred, to be so well disciplined that he will not break from a fast trot, however goaded by his driver and his own ambition in a sharp contest, into a pace in which he is conscious he can make better time with far more ease than in the one he is forced to take.

The trot appears from our analysis not to have been designed as the fastest gait, but for the medium one between the run and the walk, and when not urged too far beyond his supports ; it is the strong business gait, in which he is capable of travelling farther in a day's journey with less fatigue than any other. It is owing to this fact that it has become the favorite pace in America, and has been cultivated to a greater extent than in any other country ; indeed, we fail to learn anything of the trotting horse from any source before the importation of " Messenger," who was a thoroughbred running horse, and manifested extraordinary speed in the trotting pace after his arrival in America. It is very difficult to discover wherein the mechanical proportions and points for a fine runner would not apply equally well to a fast trotter; and it is claimed that there has been no fast trotter who did not trace his pedigree to thoroughbred ancestry. This question, however, is beyond the pale of this essay, and has not been one that has particularly interested us. It will be difficult for one to believe that a new function has been developed in the time that has elapsed since " Messenger's " importation, not yet a century, even by the most advanced Darwinian. It is much more reasonable to believe that, while great attention has been paid to breeding those qualities that insure speed, equally great care has been bestowed upon training, so that the fast horse shall display his powers in the trot rather than in the run, of which he may be equally capable. In this way the fast trot becomes a habit with the individual, and in it he may excel his powers in the superior gait. But the habit is not inherited, nor can it be transmitted to descendants, as is asserted by some authors. Functions and faculties, or the power by which acts are performed and habits acquired, may be inherited. No man is able to transmit to his offspring his acquirements, whether of mind or body. The child of the most profound scholar will not know one letter from another until he is taught them, and will learn them no more easily for all his parent's attainments. *Nascitur, non fit* (born, not made) is as true in this sense now as ever it was. The faculty through which the parent was enabled to acquire any accomplishment, whether mental or physical, may be transmitted to offspring. Even these are not congenital or coincident with birth, as the function of breathing is; but the tendency is inherited, and the functions will be developed at the proper time, and in the order that their exercise will be required, first for the existence, protection, and development of the individual, then for the full employment of all the powers successively with which it was endowed by inheritance. The law of inherited or constitutional disease is the same; all are not congenital, but most of them are developed, like consumption, at the age of puberty, and others, like cancer, at mature life, from inherited *tendencies.*

When the colt is seen soon after birth, he must be helped upon his feet, and the first efforts of his long and feeble limbs are to walk, in which he instinctively obeys the law to alternate the limbs and so preserve its balance. More than this he cannot do. Visit him a few days later, and he will be found not only able to walk with firmness, but to trot away from your approach. When you next visit him, after a longer interval of time, he has acquired much greater control of his locomotive organs, and he will move off in a trot with no uncertain step, and, if you pursue, he will break into a run (Plate XLII.).

The last pace is no less intuitive than the first, but required a longer period for its development. It is an acquired pace, but not the less intuitive. There is great doubt whether the complicated movement of the run, which has so long eluded the comprehension of man, was ever any better understood by the most sagacious of the quadrupeds that practised it; and the identical character of the movement through so many species of them shows that it is inherited, or

natural. The walk, trot, and run are all equally natural, and each is
best adapted to each of the three degrees of speed which the animal
finds it convenient or necessary to employ in his feral or unbroken
state.

It will be seen that the theory of the trot is the same as that in the
run, namely, that *the centre of gravity shall be supported constantly
and propulsion made uniformly by all the extremities from the time
they reach the ground until they leave it*, but by two alternate limbs
at a time, and not by one as in the run.   The action of the limbs in
shortening and extending, to enable them to begin the support early
and continue it late, and permit the centre of gravity to pass over them
without being deflected, is the same in both paces. The action is
the same, differing only in degree. The undulations are greatest in
the slow trot, and diminish as the speed is increased. Every rider
knows this from experience ; the uncomfortable trot is the slow one.
The reason for this was explained when treating of the run in the last
chapter.   In the slow trot the action of the muscles is not sustained,
and the bony levers are allowed to resume their normal angles.   At
each half-stride the centre of gravity regains nearly, if not quite, its
elevation ; but as the horse increases his speed he lowers the centre of
gravity, and, in so doing, enables the extremities to reach farther and
sustain the weight longer, while the rapidity of the movement of the
body gives it a momentum that forces the suspensory ligament to yield
and the angles to close to the requisite degree to prevent the alternative
of deflection of the trajectory, or crushing of the limb ; and if measurements be taken of the height of the horse at different portions of
the stride, it will be found that it is least when it would seem that
it should be the greatest, that is, when it passes the perpendicular, or
that point where the supporting limbs are shortest, as was shown in
the last chapter when analyzing the action in the run.

While the action of the limbs in the two paces is similar, the co-
ordination of them presents some interesting points for consideration.
Instead of the great impulse being given by the fore leg, as in the
run, it acts in a lesser degree, raising the centre of gravity only so far
as to give its co-operating posterior extremity an opportunity to use its

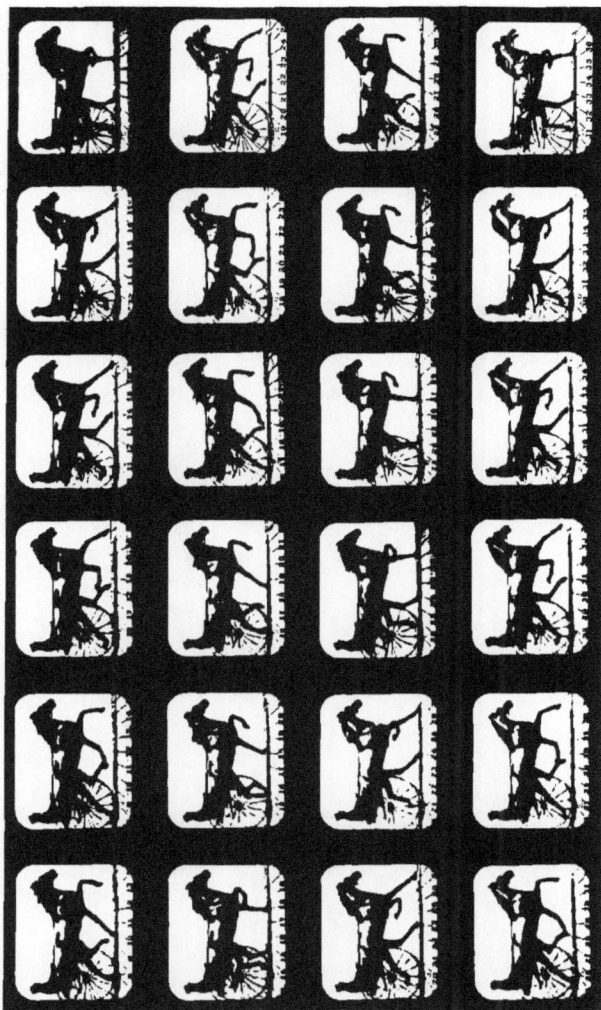

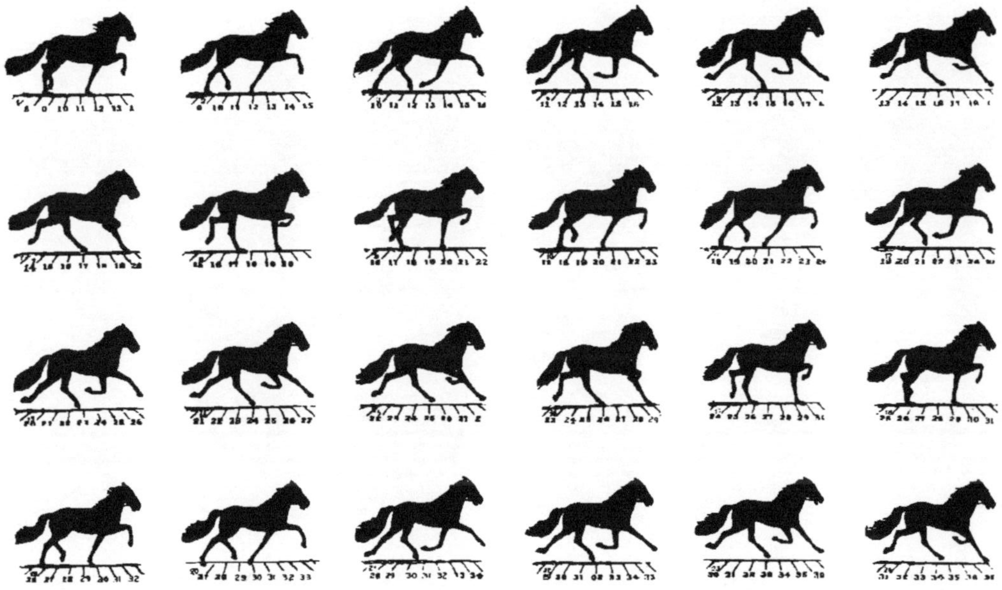

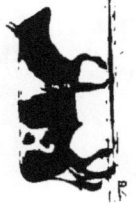

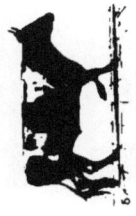

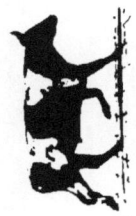

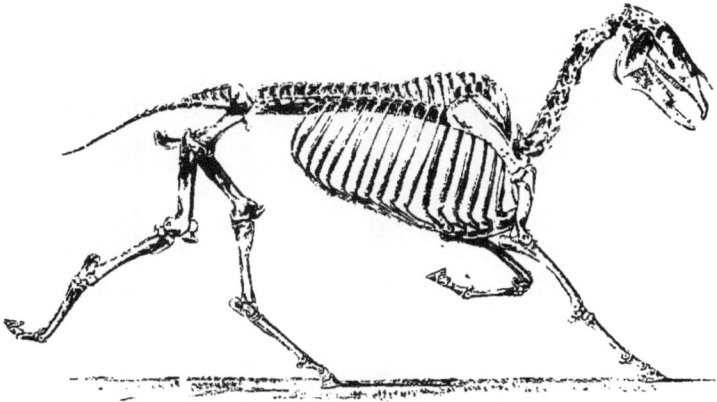

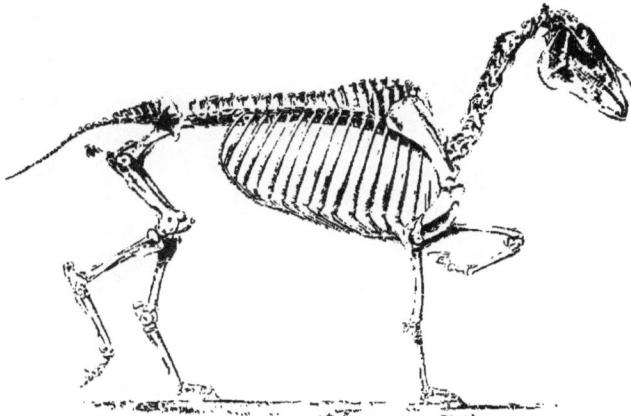

propelling power in the direction of the centre of gravity, as in the bound of the deer and the leap of the horse; for if it were given above it the animal would be pitched headlong. The hind foot should be the last to leave the ground on *a priori* reasoning ; and on consulting the silhouettes of the trotting horse (Plates XXXVI. – XL.), such will be found to be the fact. The early start of the fore foot enables it to clear the way for the hind one on the same side to advance to the support of the centre of gravity in its turn without being hit by it, or overreached, as it is technically called. In the mean time the fore foot, having a more circuitous route to travel, is enabled to attain its position as a supporter at the same instant as its co-operating partner.

The position in which the feet fall is as nearly on a line as is possible without their interference, both in the trot and run, as may be seen by referring to the illustrations of the paces in the latter part of this volume, where the animals are seen in various aspects of the same position.

Interference with the posterior feet is rendered very difficult by the mechanical arrangement of the hock joint, already explained, which causes an involuntary circumduction of the hind feet as they pass each other, and yet compels the feet to be planted successively near the same base line. In the case of the anterior extremities there is no corresponding contrivance, but the breadth of the shoulders renders it unnecessary.

For obvious reasons it is not possible to show by the camera the occurrence of interference, but overreaching is shown in Plate XXXVI. Figs. 2, 10, and in Plate XXXIX. Fig. 9. The fore foot being dilatory, or the disproportion in the length of the body to that of the legs, exposes the fetlock and heel to injury from the shoe of the hind foot; but generally the hind foot is pushed under the forward one as the latter rises.

There is a pace closely allied to the trot, and differing from it only in one particular, and that is that the limbs do not move diagonally in pairs, but those on the same side move together. This pace is shown in Plate XLIX. Comparing this series with a trotting series

15

(Plate XXXVII.), it will be found to be impossible to distinguish one pace from the other, as shown in the silhouette. This pace is called *racking*, or *pacing*, in America, and *ambling* in England. The objection to the name *pacing* is that the word *pace* is used constantly as a general term for all the different modes of progression, and therefore leads to ambiguity. While in the trot the centre of gravity falls near the intersection of the two straight lines drawn through the diagonal footprints, in the amble it is shifted from side to side, as the right or left feet alternately support the weight. The effect of this is to give a rolling motion to the body like that of a ship with the wind abeam. It is an easy pace for the rider, being free from the sharp undulations of the trot. The necessity which exists of rapidly changing the base of support from side to side makes it practicable in the horse only when the speed is considerable, and quite impossible in the rate pursued in the walk. In the camelopard, owing to the shortness of his body and the great length of his legs, it is the only method of locomotion possible, as he would overreach in the paces used by the horse. He is able to make progressive motion in this way at whatever rate, from the great elevation of the centre of gravity, and the consequent slow oscillation of it; for the time of its oscillation increases with the length of the line from the centre of gravity to the base of support.

The amble is natural to some horses, which take to it instead of the trot; as some people are sinistrous, though the greater number are dexterous instinctively, and others are ambidexterous.

Some horses are amblers first, and afterwards learn to trot and travel equally well in both paces; indeed, considering the small proportion of horses that fall into this pace, and the record made by them on the turf, it may be thought to have no disadvantage over the regular trot. It would seem to give great advantage to a short-bodied horse, as there is no danger of overreaching.

Many of the photographs reproduced in the photolithographs, and used in this volume to analyze the paces, were imperfect in lights and shades, and others, when the subjects were dark-colored, were in silhouette, to which all were reduced. The outlines are quite perfect, and the details in other respects are quite unimportant to the study of the

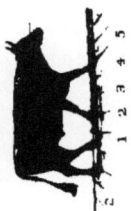

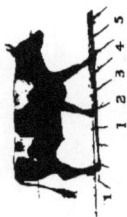

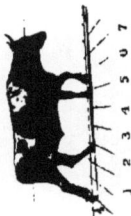

PLATE 1

movements. It is only necessary, in order to determine the movements of the several limbs, to suppose either of them to be right or left, and follow it as such throughout the stride. Examples have been selected of the two principal paces, when the horses were light-colored, to reproduce in heliotype, — a process which furnishes an exact transcript of the original photograph by the same agency, namely, the sun.

The walk is the simplest of the paces, and best understood. It is defined to be that pace in which one foot is not raised until its fellow is upon the ground. The definition is as applicable to quadrupeds as to bipeds, if in the former we assume the two anterior and the two posterior extremities as pairs. The slow walk, or saunter, represents the pendulum of writers on animal mechanics, by whom the leg was supposed to swing like a pendulum on its centre, but little muscular force being used except to counteract the attraction of gravity.

A man in walking throws the centre of gravity over the leg, which is to serve, for the moment, as a column of support, and leans forward until the centre of gravity is in advance of the foot as a base ; this renders necessary the advance of the other foot to serve as a new base, and the action of the flexor muscles upon the toes, with the weight of the suspended leg, carries the centre of gravity diagonally forward until it is again supported by the other foot. These movements are all detailed and formulated by the old writers, and are referred to here for the purpose of bringing the science of animal dynamics, as it has been taught until now, freshly to the mind of the reader.

It must be conceded that we have advanced the science of animal mechanics somewhat in this treatise, and demonstrated the fact that its problems are not to be solved by physics, as heretofore attempted, nor yet by vital force exclusively ; that animal motion in its highest manifestation is the resultant of both, chiefly of vital force, but neither can be ignored by one who would understand the subject.

Each one of these elementary acts of progression is a step, and a series of them is a walk. The walk of a quadruped is more complex and perfect than that of a biped ; for while the latter is compelled to oscillate his body in order to balance it upon each foot alternately, the

quadruped uses the diagonal feet alternately, so that the centre of
gravity always falls within the quadrangle formed by them, and near
the intersection of the lines connecting their diagonal feet.

The *theory* of the walk in quadrupeds is that there should be
two feet always upon the ground while the diagonal ones are being
advanced, and if the legs moved synchronously in pairs, there must be
four on the ground for a brief time at each step, — for from the definition of the walk one foot does not rise until the other is upon the
ground; — it follows that in two pairs of feet the two feet cannot rise
until the other two are upon the ground. This, one would think,
should be proved by the camera; but it shows that sometimes three
feet are on the ground, but never four at the same time. How is
this? Is the definition of the walk incorrect? It is so when applied
to quadrupeds. In fact, the diagonal limbs do not act synchronously
in the slow movements of the walk, for it is more difficult to maintain
an equilibrium in a slow movement than a fast, — as a top falls when
its revolutions are slow, — and for the reason that a horse never rests
on two legs, but always on the two anterior and one posterior, so
that the centre of gravity always falls within a triangle; so in the
walk one of the reserve feet holds the ground for a brief time until
the other has the start, in order to shorten the time in which the
centre of gravity has but two points of support.

The walk, being the slowest pace of the horse, has been best
observed and most discussed, but chiefly as to the order in which the
feet are moved. There can be little doubt that habit in the horse, as
in man, determines which foot shall be the first to move; and it may
often be determined by their accidental relation to each other at the
instant that he has occasion to move one of them, though it would
be doing no injustice to the brute to suppose him to have a sufficient freedom of will to choose which foot he should put forward
if he waited to think of it.

When the horse quickens his walk, he does not at once change his
pace, but extends his strides and makes them more uniform, until
further extension becomes difficult, when he will break into a trot, in
which there are never more than two feet upon the ground at a time,

as has been already stated. This change from a walk to a trot is shown in the fine silhouette (Plate XLI.).

Single-foot is an irregular pace, rather rare, and distinguished by the posterior extremities moving in the order of the fast walk and the anterior ones in that of a slow trot. These mixed paces are quite compatible, as they are of the same kind and move in the same diagonal order. It is illustrated by Plate LV. The rhythm of the footfalls is characteristic, and once heard will ever after be recognized, even in the dark. The same horse is made to illustrate the regular walk in Plate L.

CHAPTER VII.

THE series of plates which follow are intended to show more fully
than was possible in the silhouettes that precede them, the action of
the horse in every possible position in all the paces; they require,
however, a brief explanation.

The same ground was used as that on which all the experiments
were made that are detailed in the Appendix; but instead of a full
battery of twenty-four cameras, only five were employed, and they were
arranged in the manner shown in Plate I. (frontispiece). One only
represented the battery, and that was in the middle of the series; the
other four were placed at nearly equal distances, two on each side, so
as to represent the arc of a circle whose centre should be occupied
by the horse at the moment he appeared opposite the central of the
five cameras. At this point a thread was drawn across the track
which, when the breast of the horse came in contact with it, made
magnetic communication with all five of the cameras at the same
instant, so that five views of the animal were produced at the same
time, showing him from as many different directions.

The time of exposure of the negatives was so immeasurably small
that few of the pictures taken were perfect in all the details; and as
red appears as black in the photograph, so all bay horses were without
any details of light and shade, simply as silhouettes; and even when
the horse was light or gray there would be some defect in some part
of every one of the series.

Experiments were made with various processes to reproduce them
with all their defects; but it was found that the making of the

ILLUSTRATION OF THE ATTITUDE

ILLUSTRATION. — THE LADIES' LEAP.

ILLUSTRATIONS OF

ILLUSTRATION

ILLUSTRATION N°   111

necessary transfers from the originals, while they reproduced accurately all the defects of the original photographs, reproduced them with diminished sharpness, and these methods were abandoned. Under the direction of the Heliotype Printing Company another plan was adopted. From the original photographs, by the heliotype process, copies were produced on gelatine magnified, and prints were taken on Bristol board in blue ink in the same manner as in the ordinary heliotype process. These prints, with the originals, were put into the hands of artists skilled in drawing on wood for engravers, who drew them with a pen in india ink, under careful supervision of the writer, so as to preserve the outlines as they were rendered by the camera and avoid reproducing the blotted defects of the originals. These drawings were then reproduced on stone by the camera, reduced to their original size, and the prints given in the volume were printed from these stones as in ordinary lithography.

They cannot fail to be of great advantage to artists, especially those who would perfect themselves in animal drawing, and that acknowledged difficult branch of their art, — animals in motion.

They and the public generally are greatly indebted to Mr. Stanford for the enlightened liberality with which he has pursued this costly investigation, and given its results to the public without any prospect of pecuniary advantage to himself.

It will be observed that some of these pictures are so nearly alike that at a superficial view they appear the same ; but it is almost impossible that the times in which any two should be photographed should coincide, and there will be found no two exactly alike ; and the near approach to the same posture proves the universality of the law in which all the paces are performed.

In some of these plates there are but four pictures ; the fifth, owing to some serious defect or failure of the apparatus altogether, is wanting.

Plate LVII. represents a position in the run corresponding with that in Fig. 11, page 95, differing only in the fact that the right fore leg is performing its functions rather than the left, as in the cut. From this extremity the body will be projected from the ground,

and the diagonal hind is advancing to the support of the centre of
gravity. Comparing this with Plate LXV., in which one figure is
wanting, the correspondence will be found so close that at first sight
it is difficult to convince one's self that they are not identical pictures; but on careful inspection it will be perceived that in the
quartette the body is less advanced and the supporting leg is farther
from the perpendicular. The missing picture should be the first
in the regular order.

Comparing again this plate with Plate CII., the body of the horse
will be found to have advanced from the position in the former until
the supporting leg is quite perpendicular, and the other limbs are
relatively advanced.

In Plate CIV. there is still further advance; the foot is under
the centre of gravity, and the posterior extremities are being gathered under the body in the order with which they will successively
take their turn.

Plate LXXVIII. exhibits the same movement on the instant that
the propulsive effort of the limb is concluded and the foot is leaving the ground. From this last position there is an interval of one
fifth of a stride, in which there is no support given to the weight
of the body, but it is moving as a projectile until the diagonal hind
foot reaches the ground, which it is about to do in the following
plate. The left hind foot will be the first to make the contact, from
which we know that the right fore foot was the one by which the
body had been projected into the air; the right hind foot will follow
and take the ground a step farther in advance. This plate may be
compared with LXX., in which the right feet are in corresponding
positions with the left, as seen in the former. Plate XC. represents
the horse in a similar position.

The slow trot is shown in Plate LIX., and is not distinguishable
from the fast walk, as seen in the succeeding plate; it is only when
the instant of exposure of the sensitive plate of the camera is coincident with that in which all the feet are off the ground that the
walk can be distinguished from the slow trot.

Plate LXI. is also an attitude of the trot. but it is recognized by the higher action of the free limbs, and this action indicates a higher rate of speed than is possible in the walk.

In the succeeding plate the walk is again represented and is unmistakable, as the three feet are supporting weight, as indicated both by their position and the yielding of the pasterns.

In Plate LXIII. we see the sluggish run in which the speed or momentum of the horse does not permit the propulsion of the fore leg to carry the body clear of the ground before the hind ones come to the support of the centre of gravity prematurely, and which constitutes the pace known as the canter. (See page 103.)

The fast trot is shown in Plate LXIV. Plate LXVI. seems to be a fast walk, in which the groom is urging the horse into a trot. The position may be interpreted into either a walk or a trot.

Plate LXVII. represents a position in the leap, and is fully explained in the sixth chapter.

The walk is further illustrated in the two following plates.

In Plate LXXI. a position in the trot is shown where the feet are all clear of the ground. Before the fore leg, which is extending forward to reach the ground, makes the contact, it must be straightened and the toes raised, as in Plate LXIV. As already stated, it is difficult in some of the "Illustrations" to determine a slow trot from a fast walk, for there may be an instant of time in the trot when three feet are on the ground. The mechanical action is the same in both paces, and the distinction is based on the speed. This difficulty could not occur where the reader has the advantage of a consecutive series of views, as is shown in Plate L.

The heavy Clydesdale in Plate LXXII. is shown in the ambling pace in which the weight of the body is borne and the propulsion performed by the two extremities of the same side.

The canter is illustrated in Plate LXXVIII. The support is here given by the left fore leg, and the greater flexion of the diagonal right indicates that it is the next in order to perform that function. The degree of action indicates a low rate of speed, which could be attained in the trot with greater ease to the horse if not to his rider.

Plate LXXXI. represents the animal in the greatest degree of extension he reaches in the run. The posterior extremities have successively performed their functions as supporters and propellers, the anterior limbs are extended to relieve them, and for the instant the diagonal feet are upon the ground, but it is only for an instant; the weight of the body is already on the fore leg, and the only propulsive force left in the hind one is derived from the reaction of the suspensory ligament and its reinforcing tendons. This position nearly corresponds with that in Fig. 8, page 93, though a little in advance of it.

Plate LXXXV. illustrates the run in the position shown in Fig. 10, page 95. The fore leg must be straight from the elbow to the foot when it makes contact with the ground, as only in that relation of the bones forming the columns of support could the weight suddenly thrown upon them be borne. A moment's consideration of the mechanical construction of the knee-joint will suffice to convince one of this, and a weakness at that point which renders the animal liable to stumble is a very serious defect, and where it exists it indicates the loss of the balance of power between the flexors and extensors of the foot. This inflexible position of the knee-joint will be found to be universal in all the paces when the limb is sustaining weight.

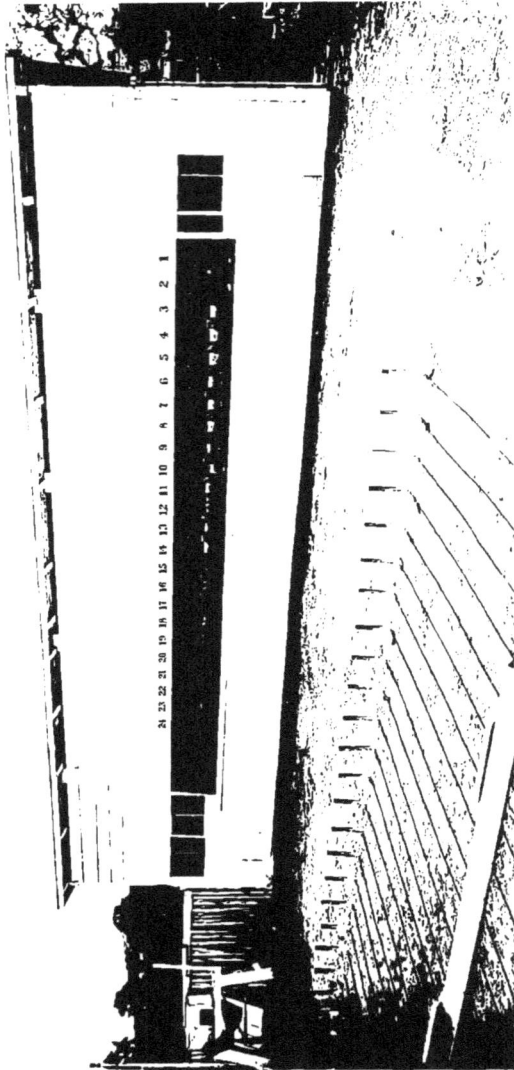

APPENDIX.

— ·——

THE following account of the methods by which the original photographs were produced that served as the basis of the analysis of the paces, the results of which are contained in this volume, was furnished by Mr. E. J. Muybridge, the photographer by whom they were executed.

Some time in 1872 Mr. Stanford, being desirous of settling some controverted questions as to the action of the trotting horse, conceived the idea that the camera might be made available for that purpose. To this end he consulted with Mr. Muybridge, and induced him to undertake some experiments in instantaneous photography.* The experiments made at that time were inconclusive, and for several years, Mr. Muybridge being absent from the State, the matter rested, though it was not abandoned by Mr. Stanford.

In 1877, Mr. Muybridge having returned, the experiments were renewed. A few pictures were taken of "Occident" while in motion — a noted trotter, owned by Mr. Stanford — with a single camera; and one of these, representing him with all his feet clear of the ground, was enlarged, retouched, and distributed among various parties interested.

* Instantaneous pictures were defined to be, at that time, in ordinary photographic parlance, when the exposure has been very brief, or under half a second. In the British Journal Photographic Almanack for 1868 it is stated that good street views had been taken in a twelfth part of a second. The conditions, as there given, for extreme rapidity of exposure are a good and quick-acting shutter, a lens with a large angular aperture, and chemicals in perfect condition.

The result of this experiment was so successful that Mr. Stanford determined to try another one on a more extended scale. He assumed, if one picture could be taken instantaneously, why not an indefinite number, and by increasing the number of cameras increase to the same extent the number of views, and illustrate the various positions in an entire stride?

Mr. Muybridge was authorized to procure the needed apparatus, and a building suitable to the purpose was erected on the west side of Mr. Stanford's private track at Palo Alto (see frontispiece). In the following year, 1878, the preparations were complete; every resource of the photographic art had been provided that was thought to be required or attainable. Twelve cameras were placed in the building at intervals of twenty-one inches, with double shutters to each. These shutters were arranged, one above and the other below the opening through which light was admitted to the lens, and held by india-rubber springs, constructed in the form of a ring, with a lifting power of one hundred pounds, and secured by latches, to be liberated on the completion of a magnetic current.*

For the purpose of making the exposures at the proper intervals of time, a machine was constructed on the principle of a Swiss music-box, having a cylinder with a row of twelve pins arranged spirally. This was put in motion by a spring, and, as it revolved, each pin in succession established a magnetic circuit, with the magnet connected with each of the twelve cameras in succession, and the whole series of exposures was made in the time occupied by a single complete stride of the horse.

---

* This description of the shutters and their mode of action is somewhat obscure. The shutter, as described by Kleffel (Handbuch der Practischen Photographie, Leipzig, 1874, p. 201), is as nearly as possible the double shutter used by Muybridge. Kleffel's shutter was held by a spring, and when the picture was to be taken the spring was touched, and the shutter, which had an opening through its centre, dropped past the lens, exposing the lens to the light during the time of the passage of the opening across it. He recommended weights to be used when greater rapidity was required. Muybridge's modification of this consisted in the use of rubber springs in lieu of weights as recommended by Kleffel; though no claim is set up by him to priority in the use of rubber springs, as one Thomas Skaife obtained a patent in England for rubber springs for camera shutters as early as 1856.

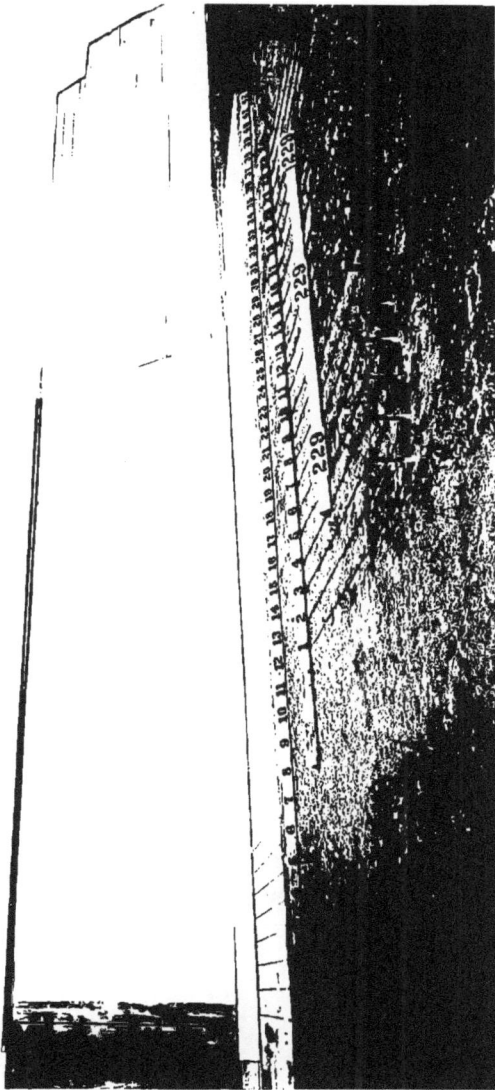

This arrangement gave the attitude of the horse as he arrived before each of the cameras in succession at the instant of exposure of the negatives. In practice it was found to be extremely difficult to set the apparatus in motion at the exact time required, and to regulate it to correspond to the speed of the horse.

This contrivance was found to be best adapted to the more irregular movements of other animals, as the running of dogs, the flight of birds, feats of acrobats, etc. It was desirable to find some method that would better represent the regular movements of the horse, and which should be regulated by his own movements.

On the side of the track opposite the building where the cameras were placed, and in such position as to receive the best exposure to light, a wooden frame was erected, about fifty feet long and fifteen high, at a suitable angle, and covered with white cotton sheeting (Plate CVII.), divided by vertical lines into spaces of twenty-one inches, each space being consecutively numbered. Eighteen inches in front of this background was placed a base-board twelve inches high, and on which were drawn longitudinal lines four inches apart. In front of this base-board a strip of wood was fastened to the ground, upon the top of which wires were secured at an elevation of about an inch above the ground and extending across the track. The wire was exposed in a groove to one only of the wheels of the sulky, being protected from contact with the horse's feet and the other wheel. Each wire was held in proper tension by a spring on the back of the base-board, so arranged that when the wire crossing the track was depressed by the wheel it should draw upon the spring connected with it, and make contact with a metallic button and complete the electric circuit.

These wires were placed at distances from each other corresponding with the cameras on the opposite side of the track, and with the spaces between the lines drawn on the background.

From this description it will be readily seen that the depression of the first wire would complete the circuit and cause the magnet connected with the corresponding camera to move the latch and liberate the shutters, exposing the sensitive plate for a space of time that is hardly conceivable. In like manner, as the wheel passed over the second wire, the shutters would be liberated on the second camera, and so on until the whole series were discharged. When the horse passed with great velocity over the wires these shutters were discharged with such force and rapidity that the horse was not unfrequently startled and broke his gait.

If everything was properly arranged the driver had but to keep the wheel of his sulky in the groove which was sunken for it, and it would, by depressing the wires successively, take the pictures at every twenty-one inches until the whole series were taken.

The method just described was used in all cases where horses were driven to sulkies; but when wheels were not used this arrangement with wires under the track had to be modified, and a thread was drawn across sufficiently high to come in contact with the horse's breast, and strong enough to cause the contact and establish the circuit as before, but not so strong as to wound the horse when going at full speed.

By these methods many views were taken and distributed to all parts of the country: they attracted a great deal of attention, and elicited a great variety of opinions and not a little ridicule; some artistic persons displayed great ingenuity in burlesque, — *no one understood them.*

The number of cameras was afterwards doubled, and they were placed at intervals of twelve inches to still closer analyze the movements of the horse. Lines were drawn across the track at corresponding distances, and the numbers indicating them, instead of being at the base of the screen, were on a board between the horse and the cameras. The heliotype plates Nos. CVI. and CVII. represent the battery of cameras and the screen as they were when twenty-four cameras were in position.

The whole of the series of twenty-four figures each used in this volume to illustrate the paces were taken in this manner. They were very accurately taken, and are specimens of the best results attained after years of expensive experience; and the heliotypes are perfect transcripts of the original photographs.

It will readily be understood that the accuracy of these analyses depends upon the uniform tension and strength of the threads connected with the springs through which the circuit is formed. The perfection of the pictures depends upon the sensitiveness of the chemicals and the time occupied in their exposure to light. This time is as nearly instantaneous as can well be conceived. Mr. Muybridge estimates it by comparing the enlargement of the horizontal diameter of an object photographed with the vertical diameter of the same object at one five-thousandth of a second. This can only be determined by measurement, and that approximately even in objects of considerable size; it is so nearly instantaneous that there is no appreciable loss of proportions from differences between vertical and horizontal diameters.

University Press: John Wilson & Son, Cambridge.